Market Study of Safety-Related Computer Controlled Systems

A review for the Department of Trade and Industry

In June 1990, the Safety-Related Systems Working Group (SRS-WG) of the Government Interdepartmental Committee on Software Engineering (ICSE) published the companion consultation documents:

SafeIT1 — ''The Safety of Programmable Electronic Systems: a government consultation document on activities to promote the safety of computer controlled systems'' and

SafeIT2 — ''A framework for safety standards''.

The formation of SRS-WG arose from an earlier recognition of the need to coordinate Government plans and activities relating to safety related software. This is important at a time when computer-controlled systems form one of the fastest growing sectors in the world's advanced economies; and where growth is being matched by developments in elaboration and sophistication of systems. Unreliability or failure in these circumstances hold serious consequences including danger to life and health.

It was recognised that the initiatives identified and published in these earlier consultation documents needed to be placed in context with up-to-date market information and that a study would be necessary. DTI therefore commissioned such a study to determine the characteristics of the safety-related computer controlled systems market and its findings are given in this report. DTI gratefully acknowledges the contributions and comments received from users, suppliers, researchers and professional organisations made during the conduct of the review.

To contribute towards success of the SafeIT initiative and to influence future work directions it is desirable that Government receives views from interested parties including users, suppliers, professional institutions and Government departments. You are invited, therefore, to send written comments to:

SRS-Working Group
Department of Trade and Industry
IMT6F—Desk 4.044
151 Buckingham Palace Road
London SW1W 9SS

Department of Trade and Industry

Safety Related Computer Controlled Systems Market Study

A review for the Department of Trade and Industry

by Coopers & Lybrand

in association with

SRD-AEA Technology and Benchmark Research

LONDON: HMSO

First published 1992

ISBN 0 11 515315 2

The views expressed in this report are those of the authors and not necessarily those of the Department of Trade and Industry

Produced from camera ready copy supplied by the DTI

Market Study of Safety-Related Computer Controlled Systems

Contents

Appendices

Executive Summary

Introduction

1 Coopers & Lybrand, in association with SRD (AEA Technology) and Benchmark Research Limited, was commissioned by the DTI to undertake a study of the characteristics of the market for safety-related computer controlled systems (SRCCS). The purpose of the study was to describe how the market was operating, to provide quantification of its key features and to identify any impediments to the effective operation of the market.

Approach and method

2 Five methods of market assessment were deployed, it being recognised as important to validate the findings from one source by cross-reference to others. The five methods involved:

- desk-research and review of the existing (mostly technical) literature;
- an interview programme with 64 organisations comprising system developers and manufacturers, software companies, engineering contractors, trade associations, training and research bodies and regulatory authorities;
- a telephone survey of nearly 500 small and large businesses in transport, energy, process manufacturing, and discrete manufacturing;
- a postal survey of over 300 industry and trade associations across Europe, of which only 40 responses were received; and
- five focus groups held after the other surveys had been completed and analysed.

3 The characterisation of the SRCCS market derived from the above methods is summarised under the headings below.

Market size and trends

4 There were two categories of SRCCS which emerged from the study - control systems whose safety related features are linked with other control aspects and specialist systems where safety is the only or primary purpose. In fact, the first category does not represent a well-defined SRCCS market. Safety is not generally marketed by the suppliers nor sought by customers but is integral with other control purposes.

5 The second category - specialist SRCCS - also does not represent a single market but rather a combination of niche markets with distinct properties and with few suppliers straddling the different niches.

6 The estimated UK market demand for both categories of SRCCS in 1991-92 was of the order of £350 million (£90 million for specialist SRCCS). The specialist SRCCS

were used to a far greater extent in transport and increasingly in energy and process manufacturing but only to a limited extent in discrete manufacture. The expectation in all sectors prior to the prolonged recession was for a significant increase in the number of both general and specialist SRCCS installed in 1992. The number of systems installed in 1991 was over double that in 1987.

7 The estimates of the SRCCS market are the first of their kind and some important qualifications need to be attached to them:

- there is no agreed definition of the market for SRCCS amongst suppliers;
- about three-quarters of those suppliers interviewed could not provide any evidence on the size and growth of SRCCS markets in the UK;
- the investment represented by an SRCCS is very lumpy, system costs varying from £½M to over £2M within the same area of application - so small variations in installation numbers can have significant but unpredictable effects on the value estimates of market size and growth.

8 If the growth in the Western European market of SRCCS installations was much the same as estimated for the UK (about 25% per annum over the late 1980s) then one estimate of the market size would put it at £1000 million in the early 1990s. This would imply a UK market share of about 35%. Even though it was claimed by suppliers that the UK was well in advance of other European countries in the off-shore market for SRCCS and perhaps also in petrochemicals, nevertheless this market share seems high and suggests the European market size may be under-estimated.

Hazard and risk assessment

9 In broad terms, SRCCS were used:

- in the transport sector to protect the general public against injury arising from a high frequency probability of process malfunction;
- in the energy sector to protect against a wide range of frequency probabilities of process malfunction, plant and environmental damage;
- in process manufacturing to protect operators and the general workforce against injury arising from a low frequency probability of malfunction; and
- in discrete manufacturing to protect operators against injury arising from the range of frequency probabilities of malfunction.

10 The decision to opt for an SRCCS solution was driven by the imperative to minimise risk and comply with regulations. Although cost-benefit assessment was a second-order consideration, nevertheless 70% of the users of SRCCS provided training in hazard assessment. However, only 40% of users trained in system specification and the suppliers claimed that requirement and system specification was the weakest link in the safety system life cycle and the most fundamental.

Strategies for risk reduction

11 Strategies for risk reduction using SRCCS were likely to have been inadequately thought through because:

- there was a significant degree of ignorance about strategic SRCCS options amongst potential users;
- formal hazard assessment and cost-benefit analysis were not used to any great extent in system specification and design;
- there was a general tendency to use in-house resources for system specification, design and validation but there was limited training provision on these matters.

12 There is no evidence of these factors having an adverse impact on SRCCS performance. But they imply that users are not generally making fully informed choices in their risk reduction strategies.

Techniques and standards for safe systems

13 There is widespread recognition amongst suppliers that standards are required but:

- there is concern that standards are being driven by academics rather than industry;
- there is little support to encourage industry to participate actively in the development and formulation of standards;
- the standards tend to be procedural - they focus on the process rather than the goal of system development (DEFSTAN 00-55 was considered by many to be a useful standard because of its more 'goal orientated' approach).

14 There was less concern amongst suppliers about the sort of software used than about the way in which programs were verified and validated. As an extension to in-house verification, validation and testing, suppliers noted the attraction of third party approval - the German TÜV system was referred to in this regard.

15 Training was regarded as essential for all those involved in the design and development of systems. But this requirement was not specific to safety. Professional testing and certification was required for those working in safety system design, not training per se.

Safety in operation and maintenance

16 It is essential that there is understanding amongst users about how their actions (in operation and maintenance) could affect the parameters of the safety system. Most users provided training in system use and maintenance, but little use was made of external sources of training. This suggests that users send only limited numbers of operators and maintenance supervisors on external courses, expecting them then to pass on acquired knowledge in-house.

17 Little enthusiasm was registered by over half the users for more training on the grounds that adequate provision was already made and safety was already built into the system. This suggests complacency.

External factors and their market impact

18 Legislative and regulatory pressures and technological change were the two factors thought by users to drive industry towards improved levels of safety and use of SRCCS. However, the study suggested that increased demand for SRCCS was most unlikely to come from those currently not using SRCCS because of their lack of awareness of this option.

19 Despite these pressures, there was little evidence from the study that there would be convergence amongst the niche SRCCS markets. Close user-supplier relationships in the niches are well established and suppliers see very little need to market their products outside their existing customer base.

20 This may put UK industry at a competitive risk if European wide-standards for SRCCS are developed formally or de facto (eg through the German TÜV approval system).

Market impediments and possible areas for action

21 Impediments to the effective operation of markets may suggest areas where action might be appropriate on the part of government, industry and the infrastructure organisations. The categories of market impediments which may be of particular relevance to the SRCCS markets are:

- incomplete information and awareness amongst users and potential users about safety system options and standards which may lead to less than optimal choices of safety systems - there is a clearly expressed need for more information of a generic kind and in a co-operative context (eg through an association or club);

- market segmentation into self-enclosed niche markets (eg the various specialist SRCCS markets) which may limit broader based competition and inhibit cross-niche developments (eg in technology transfer, standards and training);

- insufficiently broad-based demand for training in SRCCS and other safety systems' specification, implementation and maintenance which may limit developments in the supply of training of potential benefit to all system users;

- limited development of generic standards and guidelines with regard to safety systems; and

- rapid technological change and the gradual development of international competition and standards which could pose a competitive threat to both suppliers and users in the UK if they remain locked into niche markets.

22 The programme of interviews with suppliers and infrastructure organisations carried out during the study drew out various observations on the initiatives thought desirable. Whilst there were quite different views on appropriate specific actions, there was general agreement on the following categories for further consideration:

- awareness generation and information provision;
- standards and guideline specification and promulgation;
- training standards and provision; and
- transfer of best practice.

Most weight was attached by those interviewed to the first two categories of possible initiatives.

1 Background and rationale for the study

Introduction

101 Over the past decade industry's use of computers, or in more general terms computer systems, has increased dramatically in a variety of applications. This reflects technological advances together with the dramatic reduction in the cost (in terms of processing power) and physical size of such systems.

102 Attention has increasingly focused on ensuring that computer systems are safe. There has been growing awareness that failure of all or part of a system could impact adversely on safety. This was well-understood in "life-critical" applications (eg in aerospace or nuclear reactors, for emergency shut-down or fire detection and extinguishing). It is also relevant where systems are used for more general control applications (eg for plant, process or machinery control) where failure of the system may result in loss of human life, environmental damage or a demand being placed on a safety back-up system.

103 The safety related attributes of systems were highlighted in a report by the Advisory Council for Applied Research and Development (ACARD) in 1986. The report indicated that system failures were primarily related to the programs which control them:

> "..... programs full of errors, oversights, inadequacies and misunderstandings of the programmers who compose them." (pp 78)

104 While software is an important component of a system, it is only one amongst many. Other elements of a system include:

- the components used in the production of these systems (eg the reliability of hardware and their susceptibility to failure); and
- the process by which these systems are produced (eg the tools and techniques used, the competency of the individuals involved and the management and control procedures).

105 Following on from the ACARD report, there have been a number of initiatives to devise a strategy to assure the safety of electronic systems. For example, in the UK, the Health and Safety Executive (HSE) published guidelines in 1987 for the safe use of programmable electronic systems (PES). This was followed by more specific interpretation and guidance for particular industry sectors, such as that produced by the National Association of Lift Makers (1990) and the Engineering Equipment and Materials Users Association (1989).

106 The Department of Trade and Industry (DTI) part funded a study in the field. The study, prepared jointly by the Institution of Electrical Engineers and the British Computer Society (1989), examined the role of software in safety-related systems. It incorporated an examination of the computer-based systems safety practices in the UK, the rest of Europe and the USA. This was undertaken by the Centre for Software Engineering.

107 In June 1990, the Government Interdepartmental Committee on Software Engineering (ICSE), published two consultation documents on the safety of computer-controlled systems (SafeIT 1 and 2). The first made recommendations to:

- facilitate the development of technically sound, feasible, generic international standards together with consistent sector of application-specific standards which are likely to achieve wide acceptance;
- encourage the use of PES technology in safety-related applications and promote the adoption of advanced techniques and best IT practices in relation to PES;
- ensure that the application of PES in safety-related applications enhances safety;
- find better technical solutions to assure PES integrity; and
- promote, within the UK and internationally, open markets in safety-related PES.

108 The second consultation document was concerned with the creation and development of a Safety Standards Framework. Its stated purpose was to provide the necessary structure to allow PES safety standards, currently being developed disparately, to be integrated.

Rationale for the study

109 The SafeIT consultation documents emphasised that the Government had the objective of promoting the safety of computer controlled systems. Problems with these systems were identified which could:

> ".... increase the costs of using PES, pose potentially unacceptable social and economic risks and jeopardise the considerable safety and cost benefits that can arise from their use." (SafeIT 1 pp 7)

110 Specifically, these problems were associated with a wider range of industry becoming dependent on PES at a time when the technology involved was still relatively immature. The systems are therefore being introduced into new areas where:

- there was a lack of understanding and awareness of the hazards and risks that need to be addressed;
- there was a lack of familiarity with the specialist techniques for the development of software, particularly for applications requiring a high level of integrity; and
- the software suppliers could be unfamiliar with basic safety engineering principles.

111 These problems were thought to be exacerbated by the sectoral development and application of PES, including the methodologies, techniques, standards and supervisory bodies. As such, there might be little cross-sectoral diversification and integration. There could be significant barriers to the adoption of best practice and the exploitation of economies of scale in systems development.

112 On the basis of these observations, the SafeIT consultation documents suggested the following possible Government initiatives:

- the development and harmonization of standards at a European and international level;
- the promotion of technology transfer between sectors;

- the requirement for collaborative and long term fundamental research and development in relevant technologies; and
- the development of a co-ordinated approach to education and training for the development of safe systems.

113 Government has subsequently taken action in most of the areas identified by the SafeIT consultation documents. However, for these initiatives to be developed in an effective way, it has been recognised that they needed to be justified on comprehensive and up-to-date market information. To date, this has not been available.

Study terms of reference

114 Based on the recommendations of the SafeIT consultation documents, the DTI commissioned Coopers & Lybrand in association with SRD (AEA Technology) and Benchmark Research Limited to undertake a study of the characteristics of the safety-related computer controlled systems market. The terms of reference set out six principal objectives for the study:

- to quantify the market for safety-related computer controlled systems throughout the European Community and the European Free Trade Area;
- to analyse the market in terms of its key characteristics;
- to describe how the market for control equipment operates, especially in terms of those features relevant to the rationale underlying SafeIT activities;
- to quantify, for the purposes of the market covered by the study, the post-experience education and training situation particularly with regard to software engineers;
- to explain how the gathered data and its analysis have been validated as representative; and
- to estimate how the market might develop over the next ten years and to identify factors which would maintain and enhance the UK position, taking account of developments in standards, technology and the market itself.

115 The detailed terms of reference for the study are set out in full in Appendix A.

116 In commissioning a study with these objectives, the DTI was seeking to establish how the market for safety-related computer controlled systems is operating in order to identify where the market may not be working adequately in protecting human life and well-being and the environment. There may then be a case for the DTI or other agencies, industry and the supporting infrastructure organisations to act to help rectify any market impediments.

117 The next section of this report provides a brief outline of the key concepts and terminology that form the basis of the study. Section III describes the approach and methodology used and Section IV provides a review and analysis of the key findings of the study. The implications of these findings for market impediments and potential areas for action are set out in Section V.

II Concepts and terminology

The need for a common set of concepts and terminology

201 The objective underlying the SafeIT consultation documents was to promote the widespread use of IT in accordance with best practice. For safety-related systems there was particular concern that they should be ".... constructed to satisfactory standards and are adequately safe." (SafeIT 1 pp 3).

202 One of the ways in which this objective could be met was through the development of a framework for safety standards. One of the key requirements for this was the development of a common set of concepts and terminology because:

- standards bodies and some industry sectors were in the process of developing, or had already developed, their own sets of terminology and concepts;
- many of the difficulties in achieving the necessary consensus for standards were associated with terminology and concepts and how they were interpreted; and
- the harmonization of standards required common concepts and terminology.

203 The consultation documents identified the importance of common concepts and terminology but were not intended to develop them in a systematic way.

A systems approach

204 A useful way of demonstrating the concepts and terminology relevant to our analysis of safety-related computer controlled systems is to use a systems approach. The principal elements of this are set out in Figure 1 which shows how we have built on the work reported in the SafeIT documentation and other material. It shows that

Figure 1 : Key Risk, Integrity and Systems Concepts

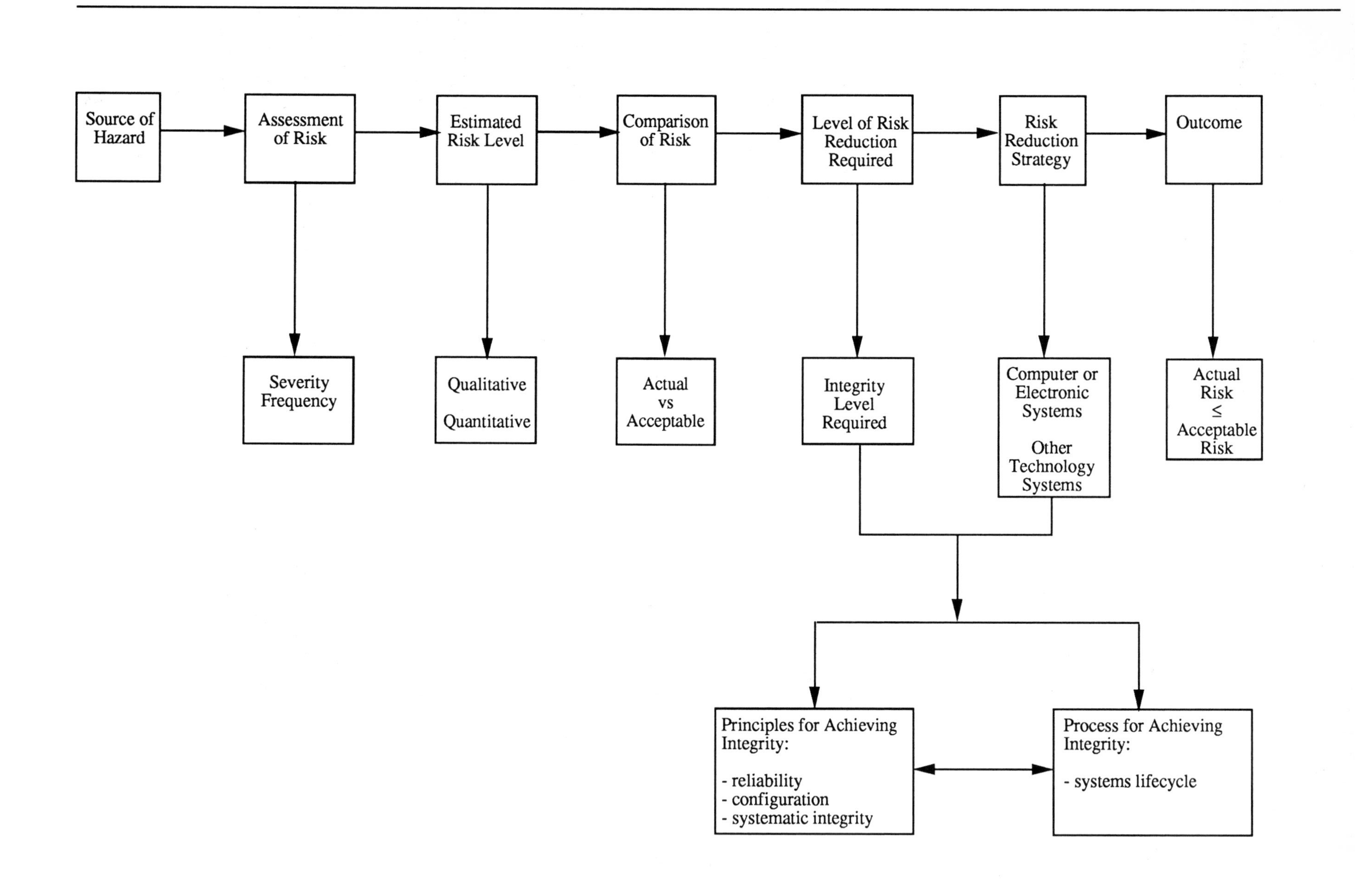

GSE-32-6

safety is an attribute of a system as a whole. Safety is defined as a situation where the risk of a hazard occurring, likely to result in human death or injury or damage to the environment, is at a level which is acceptable. The concept of acceptable risk is used since no situation will ever be entirely free of some element of risk. Acceptable risk is therefore the highest level of risk associated with a situation that is justifiable; it is a measure of 'safeness'.

205 In industry, systems usually comprise a number of elements, or sub-systems, which will all be a potential source of hazard. They include:

- housekeeping procedures;
- processes being controlled;
- materials or feedstocks used;
- environmental effects (eg vibration, heat, electromagnetic effects);
- plant and equipment under control;
- individuals involved (eg in operation, design, development, maintenance and management);
- programmable electronic control systems; and
- control systems based on other technologies (eg mechanical).

206 In systems where computer or electronic systems are used, the hazard may arise as a result of one or more of the following types of failure:

- random hardware failure - failure of hardware arising from the degradation and breakdown of hardware components and mechanisms that occur at unpredictable times;
- systemic failure - failure which can occur due to errors made at some stage of the specification, design, development, constructi[illegible]ommissioning, operation, maintenance or management of a system, a[illegible]ch becomes evident as a result of particular combinations of input[illegible]nvironmental conditions.

207 To determine the overall level of risk associated with a system, the systems approach requires an analysis of the potential sources of hazard of the system as a whole, including its component sub-systems. The level of risk is typically measured as an outcome of:

- severity - the nature and extent of the consequences of a hazard occurring; and
- frequency - the likelihood of a hazard occurring.

208 The risk associated with a system is therefore a product of these two dimensions. These may be expressed using qualitative or quantitative scales, for example:

- the severity of a hazard may be described as:
  - catastrophic: involving a large number of deaths, disabling injuries and/or extensive, environmental damage;
  - critical: involving few deaths, disabling injuries and/or more limited environmental damage;
  - marginal: involving minor injuries and/or local environmental damage; or
  - negligible: involving damage to the process, plant or product, resulting in economic loss;
- the frequency of occurrence may be described as:
  - frequent: many times per year;
  - probable: once a year;
  - occasional: once during the lifetime of a system;
  - remote: unlikely to occur but requires consideration; and
  - improbable: unlikely to occur.

209 A comparison of the estimated risk of a system with the level of risk that is acceptable indicates the reduction of risk required before a system can be regarded as safe. The level of risk that is justifiable is determined by:

- legislation;
- regulatory authorities;

- industry guidelines or standards;
- systems operators; and
- independent experts.

210 Where the estimated overall risk is higher than is considered acceptable, a strategy to reduce risk to the justifiable level is required for a system to be safe. The systems approach identifies two broad strategies to achieve the necessary reduction in risk based on:

- control techniques, using computer or electronic systems; and/or
- protection techniques, using non-computer systems (eg exclusion zones, physical barriers, hardwired relay logic and mechanical reliefs).

Both approaches to risk reduction are safety-related.

211 To assure safety, the next step in the systems approach is to identify and determine the level of integrity necessary to achieve the required reduction in risk. This involves consideration of the extent to which the chosen strategy will perform the necessary risk reduction functions and the degree of certainty that can be attached to those functions being executed when required. It provides the basis for translating safety into system engineering specifications, in terms of hardware and, where computer and electronic systems are involved, into software requirements.

212 The basic concept of safety integrity has been developed by those bodies involved in developing and setting standards. The Ministry of Defence Standard DEFSTAN 00-56 identifies four levels of integrity and the separate draft standard developed by the International Electrotechnical Commission (IEC), for software in safety-related systems, suggests there are four or five, depending on the way it is interpreted.

213 The systems approach implies that the level of safety integrity, to achieve the necessary reduction in risk, should be engineered into the system. This is made clear in the SafeIT consultation documents where it is argued that accepted methods for good design and management are prerequisites for the development of assurably safe systems,

particularly when systems are used for safety-related applications. The principles underlying good design and management involve consideration of:

- the reliability of the system to ensure that random hardware failures are mitigated;
- the configuration of the system as a whole and the safety-related system(s) which comprise it; and
- the systemic integrity to cover the actions taken to eliminate the incidence of systemic failure (ie quality assurance practices).

214 As a way of implementing these principles, a safety systems' lifecycle approach has been developed. It is similar to the development lifecycle of more conventional projects. The approach is depicted in Figure 2 which shows the key processes necessary to engineer an assurably safe system, namely:

- hazard assessment and the analysis of risk;
- requirements specification (in functional and safety terms);
- system design, implementation and verification (the verification process being used to ensure a faithful translation between specification against the design and the implementation against the design);
- system integration, installation and validation (the latter being the comparison between the expected and the actual performance of the system);
- system operation; and
- maintenance and modification (to monitor and maintain a system in a safe state and ensure that any changes are introduced in a controlled way).

Figure 2 : A Safety Systems Lifecycle Model

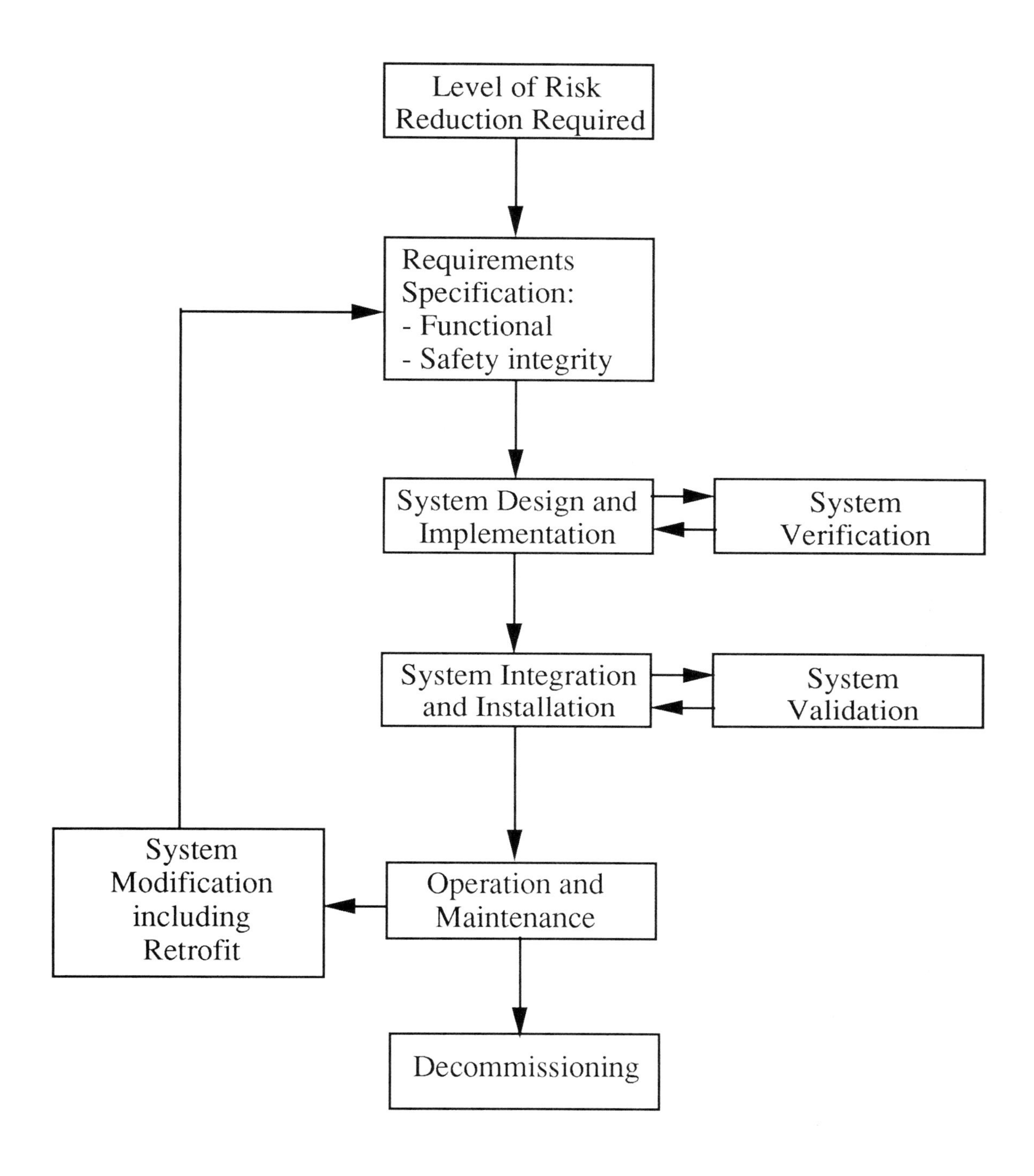

Implications for the study

215 The systems approach outlined above is 'best practice' in the development of safety-related systems. This approach provides the framework for assessing the roles and responsibilities of the parties involved and, thereby, understanding the market. At each of the key stages in the systems approach it is possible to assess the extent to which users and suppliers of safety-related computer controlled systems and other parties involved adhere to this approach and to describe the mechanisms used to implement good practice. This can then be used to provide an indication of where, in the approach, differences exist between actual and best practice, the nature of these differences, why they occur and which of the various parties are involved.

III Approach and methodology

Defining safety-related computer controlled systems

301 At the start of the study, it was recognised that the market assessment would require a clear definition of a safety-related computer controlled system. There were four particular problems which presented themselves:

(a) Safety is an attribute of a safety-related system. It is not a clearly definable product; rather it is derived from the reduction in risk that is achieved by using such a system. In practice, safety is therefore not something which can be bought and sold as a separate commodity, partly because purchasers do not expect a system to be "unsafe". This has particular implications for any form of market study, since market structures and behaviour are determined by the interaction of buyers (demand) and producers (supply) of a well-defined product.

(b) Safety can be interpreted in a number of ways, depending on whether the purpose of using a safety-related system is to reduce the risk of one or more of the following outcomes:

- (i) death or injury to people involved in the production process;
- (ii) death or injury to people not involved in the production process;
- (iii) damage to the environment;
- (iv) economic loss arising from damage to the process plant or product.

(c) There are two ways of interpreting what is a computer controlled system. The one interpretation is where a system is based on the use of electronic components. The other, more generic interpretation, is where the computer or electronic components are programmable ie software is also used.

(d) The extent to which a computer controlled system can be regarded as safety-related depends on where the boundary is drawn between a failure of the system and the ultimate impact of that failure. This, in turn, is dependent on the nature of the causal relationships involved and whether consideration is given to the primary, secondary or other downstream effects of a system failure.

302 These problems were addressed in detailed discussions with the DTI and its panel of specialist advisors. In the light of these discussions it was decided that, for the purpose of this study, a safety-related computer controlled system will have the following characteristics:

- the computer system will be programmable and be used for monitoring or control purposes;
- the primary purpose of the computer control system will be to reduce risk to an acceptable level ie it has a primary safety role; and
- the computer control system is used to protect against the risk of human death or injury or damage to the environment.

303 Given the problems with definition, considerable care was taken in industry interviews and discussions to establish that respondents clearly understood the particular systems this study sought to address. To this end, the wording used in all introductions was as follows:

> "These are systems that control and/or monitor machines and processes, with the specific remit directly to prevent both immediate and long-term danger to both the human population and the environment, including physical property."

304 This definition, particularly in terms of its scope, gave rise to considerable debate during the course of the study. Respondents often found it confusing to discuss whether such systems were safety-related. They suggested that the distinction between computer control for control and safety was blurred and, as a result, any computer control system could be considered to be safety-related. Moreover it was argued that the boundary of what is safety-related, in terms of computer control, cannot be readily delineated. It depended on how far downstream the failure of a computer control system could have an impact on safety.

305 The general conclusion was that in order to undertake any sort of market study some definition was necessary despite the difficulties. The definition used for the study, while generally accepted as valid, was considered to be rather specific in the sense that it:

- excluded the risk of economic losses; and
- focused on the direct impacts of a hazard occurring.

306 Despite this, it was admitted that, given the current level of understanding of the market for safety-related computer controlled systems, the definition used was probably the only one that was workable. If a wider definition was used, it would be unlikely to provide any real understanding of the market issues and the behaviour of those involved beyond a superficial level. Some influential, major players on the supply side challenged the possibility or usefulness of defining markets for safety related computer controlled systems other than in a few pockets of specialised work.

307 The following examples indicate the types of systems which were included or excluded for consideration in the study. Included were fire and gas detection systems based on software driven programmable electronic control. However, the same system based on hardware logic control was excluded. A software driven programmable electronic control system was included which monitors a guard switch on a milling and grinding machine and activates the cut off from the primary power source if it is compromised. A similar control system which monitors cold store temperatures and activates a refrigeration compressor to maintain the temperature at a pre-determined level was excluded.

Methodology

308 The initial focus of the work was to identify and review existing information. This comprised desk research accessing a variety of data sources. The bibliography of source material reviewed during this early stage of the study is provided in Appendix B.

309 Despite the extent of reference material obtained, its major limitation was the absence of any relevant market information. This reinforced the case for our study. The issues addressed in the material primarily related to achieving the safety and reliability of computer systems in terms of:

- the technologies available;
- the requirements, tools, techniques and competencies for producing safe systems;
- the role and difficulties associated with the development of safe software; and
- assessing existing or proposed standards in the field.

310 Nevertheless, the material and the issues it raised provided a common frame of reference for those involved in the study, which was especially relevant in view of the problems with the concepts and terminology involved. A summary of the issues arising from the desk research is provided in Appendix C.

311 The three principal methods for the collection of evidence on the market were:

(a) discussions with **the supply side** of the market, including suppliers and software companies and those parties and key influencers representing the market infrastructure and regulatory framework;

(b) a series of telephone interviews on **the demand side** of the market across different industry sectors in the UK; and

(c) a postal and telephone **survey of industry and trade associations and professional bodies across Europe**.

The supply side interviews

312 The discussions on the supply side were conducted by Coopers & Lybrand consultants. They were conducted face-to-face with senior individuals in organisations either considered to be actively involved in the area of safety-related computer control or known to be interested in the issues raised by it. In identifying potential respondents, emphasis was placed on selecting representatives of organisations which had a European or international market presence.

313 On the basis of similar market assessment studies, it was felt that being able to discuss a range of specific issues in a more informal way would elicit more accurate information on the respondent's knowledge and understanding of the market and practices of the various parties involved. However, one disadvantage of the approach used is that the information is less amenable to a quantified statistical analysis. As a result, the findings are presented in a qualitative way. The discussion guides used for the interviews are provided in Appendix D.

314 A cross section of organisations representing the supply side were approached during the middle of 1991. In total 64 interviews were completed with UK based respondents. The proportion of the sample under each of the categories representing the supply side is shown in Table 1.

Table 1: Distribution of supply side interviews

	No	%
Systems developers and manufacturers	24	38
Software companies and consultancies	16	25
Engineering contractors	7	11
Trade associations and other third parties	6	9
Training and education providers and academic research departments	7	11
Regulatory authorities	4	6
Total	64	100

315 It was originally intended to conduct a small number of interviews with various supply side parties based in Europe. A number of approaches were made but met with little success. However, UK subsidiaries of other European country companies were included in the supply-side interview programme.

316 Where interviews were conducted, respondents were both willing to cooperate and enthusiastic about the nature and scope of the study. To a certain extent, the respondents' enthusiasm reflected a desire to exchange views and to obtain additional information about the market themselves. The respondents gave the impression that there was a distinct lack of information about the safety-related computer control systems market and they looked forward to the completion of this study in order to overcome this deficiency.

317 The discussions were undertaken on the understanding that the views and information supplied would remain confidential and unattributable. As such the findings in the report have been written to comply with the assurances given.

The demand side interviews

318 The telephone interviews with user end sites were completed by Benchmark Research. This approach was used because experience of similar market surveys suggested that a postal survey usually yields a lower response rate. Moreover, given the

complexities of the SRCCS market, it was recognised to be necessary to assist respondents with their replies by leading them through a structured questionnaire over the telephone thus expediting a positive reply and often providing additional information which might not otherwise have been forthcoming. A copy of the telephone questionnaires used for the demand side survey of both users and non-users of safety-related computer controlled systems is provided in Appendix E.

319 In general, Benchmark Research reported that, once an appropriate respondent at the end site had been identified, interviews were completed in a straightforward manner. There were a number of occasions when respondents were recontacted to check on their original replies but overall the information was considered to be coherent and complete.

320 A total of 497 interviews with end sites using safety systems (not all SRCCS) in the UK were completed. The distribution of the sample by broad industry sector is presented in Table 2.

Table 2: Distribution of demand side interviews

	No	%
Manufacturing:		
- discrete (including aviation, defence and medical)	212	43
- process (including pharmaceutical)	207	42
Energy (including nuclear, gas, coal, electricity and water)	49	10
Transport control (including road, rail, air and shipping)	29	6
Total	497	100

321 A survey of this size was considered necessary to provide a statistically valid representation of the population of end sites within each of the sectors. The sample of transport sector end sites is less representative than for the other sectors because of the sample size and because some sub-sectors proved difficult to cover. The following sub-sectors within transport were contacted and completed interviews achieved: passenger airlines, private airports, railways, road transport (highway control) and shipping. Information on air-traffic control was relatively limited. A detailed analysis of the demand side survey is provided in Appendix F which also demonstrates how the survey

sample was grossed up to yield population estimates within robust confidence limits for manufacturing, energy and road and rail transport.

The European survey

322 The third element in the market survey programme was a postal survey of appropriate industry and trade associations, professional bodies and national and international organisations across Europe. This was also conducted by Benchmark Research. The postal survey was followed up by telephone reminder calls to maximise the response rate. It was our intention to obtain about eight responses in each of the countries within the European Community and the European Free Trade Area. Over 300 appropriate European bodies were contacted. Of these, only 41 responses were received and only four were of a detail and quality to be usable. The remainder were unusable either because the survey form was not completed, the respondent declined to provide any information or the organisation had moved or had been disbanded.

The focus group discussions

323 The objective of using the three-fold approach described above was to obtain as full a range of perspectives across the same range of market issues as possible. This had the advantage of providing a useful basis for validation by cross-checking the information gathered.

324 Analysis of the information obtained during the course of the market survey work provided an indication of some of the more significant issues which faced the demand and supply sides and warranted further investigation. It also highlighted where there were still information gaps once the market survey had been completed. Five focus group meetings were held, one in London, Bristol and Birmingham and two in Manchester. Approximately 65 organisations which had indicated an interest during the original supply and demand side market surveys were re-contacted and invited to attend at the venue location convenient for them. Of these, 46 confirmed their attendance. However, representatives from 26 of the 46 organisations could not attend the meetings because of business constraints.

325 While the number of respondents at the focus groups was less than expected the groups nevertheless still provided a useful way of supplementing and refining the findings that had emerged from the market survey. In addition, they provided an additional means of validating the analysis which had been undertaken by affording the opportunity to probe fully where comments and views expressed were of particular significance.

IV Characteristics of the market

Introduction

4001 This section of our report depicts the characteristics of the market for safety related computer controlled systems (SRCCS). It does so by drawing on our detailed analysis of the evidence from our interview programme with suppliers and key influencers and from our survey of users.

4002 We have adopted a presentation framework for our findings which follows closely the safe systems approach described in Section II:

- hazard analysis and risk assessment practices;
- strategies to reduce risks to tolerable levels;
- tools and techniques to produce safe systems;
- operation and maintenance procedures; and
- external factors and their impact.

4003 This framework allows a comparison to be made between what is regarded as "good" practice in the design, development and use of safety related computer control systems and what takes place generally in the market or in particular parts of it.

4004 However, the presentation of our findings in this way needs contextual evidence on the market for SRCCS. This is addressed in the next subsection.

The market for safety related computer controlled systems

4005 The view of suppliers was generally that it was the application of a computer control system which determined whether it was safety-related. A distinction was made between computer systems used for the control of process, plant and equipment (control systems) and those used to protect against the occurrence of a potential hazard (protection system). Only the protection systems were regarded as safety-related.

4006 In one sense this may seem self-evident; failure of a protection system may result in human death or injury or extensive damage to the environment; failure of a control system may cause damage to plant and equipment or loss of production. However, the distinction presumes that failures of a control system are prevented from causing death, injury or extensive environmental damage through the use of non-computer based protection systems (eg exclusion zones, physical screens or barriers, hard wired interlocks and mechanical reliefs). If these systems fail, at the same time as a failure of a control system, then this might lead to death or injury as well as damage to plant and equipment.

4007 While this possibility was recognised by the suppliers, nevertheless they worked with the more specific definition of safety-related computer controlled systems. However, within this latter definition there was a further distinction, more blurred than the first but significant in terms of market definition. This was the distinction between what might be called specialist systems and others. Specialist systems were characterised as computer controlled with the specific purpose of controlling for safety. Other SRCCS were seen to be more general control systems where safety was one amongst a number of other control attributes.

4008 The reaction in the focus groups was that the specialist definition, whilst probably too specific to capture the full extent of safety related system applications, was nevertheless likely to be the only one that was workable on the current understanding of the market. This is a fundamental point for the study since it implies that a market, in the economic sense of buyers and seller negotiating the exchange of a well-defined commodity, may only exist for specialist SRCCS.

4009 Moreover, there was a strong view from the supply side that the specialist SRCCS market should more properly be regarded as an amalgamation of niche markets with distinct properties.

4010 Slightly more than three-quarters of all those interviewed amongst systems manufacturers, software companies, R&D organisations, engineering contractors, trade associations, trainers and regulators could not provide any evidence on the size and growth of SRCCS markets in the UK. The estimates that were provided were for segments within transport applications, petrochemicals and oil and gas production. Even here the firms had very different views on the market size in these segments.

4011 The number of sites installing specialist SRCCS in each of these niche markets in any one year will tend to be low. The investment represented by such installations will also tend to be "lumpy" and consequently the demand for specialist SRCCS and/or the average cost of the systems will vary from year to year. The lumpy nature of orders for SRCCS appears to be a particular concern for the smaller more specialist systems manufacturers. Of the 24 interviews with systems manufacturers, six of them specifically highlighted the fact that the bulk of their turnover was accounted for by the supply of a small number of very high value systems. There is no reason to believe that the mix of low and high unit cost systems will remain constant from one year to the next. This may explain why some firms in the same market segment quoted annual demand estimates twice as high as others.

4012 In summary, the evidence of this study is that there are three distinct groups of safety related systems. There are:

(a) non-computer controlled safety systems;
(b) control systems whose safety related features are integrated with other control functions; and
(c) specialist SRCCS where safety is the only or primary purpose.

4013 The first group is of interest to this study as a potential source of demand for SRCCS and it provides a benchmark against which the attitudes and practices in the other two groups can be compared and contrasted. The second group is of direct relevance to the study but we conclude that it does not represent a well-defined market. There is evidence that safety is not marketed directly by the suppliers nor sought by customers but is secondary to other control purposes. The third group is closest to a market but is more an amalgam of niche markets.

4014 The survey of the demand-side (Appendix F) covered the three groups of systems, (a)-(c). Figure 3 summarises the structure of the sample surveyed and Table 3 illustrates the applications in each broad sector.

4015 The sample was structured in a way which enabled the estimation of some of the key characteristics of SRCCS in the UK, ie specialist systems (c) and more general computer controlled systems (b). These are presented in Table 4 and their derivation is shown in Section II of Appendix F. The build-up of the number of SRCCS installed is shown in Figure 4 and of the number of sites installing SRCCS in different sectors is presented in Figure 5 (which also includes the planned increase in 1992).

4016 The estimates provided here of the number of systems installed in the UK and of the value of annual demand for SRCCS are the first of their kind based on an extensive review of the demand and supply side of the markets. Consideration of the estimates needs to bear in mind the qualifications that arise from the definitional difficulties discussed earlier and the limitations of the sample in the transport sector.

4017 The following observations and conclusions from the demand-side analysis are particularly relevant to the purpose of the study:

- The number of SRCCS in use in the UK was estimated to be at least 21,350, of which some 30% could be described as specialist systems.
- The growth in numbers of sites using SRCCS was particularly marked in manufacturing, mostly in non-specialist systems.
- The expectation in all sectors was for a significant increase in the number of sites installing systems in 1992, representing a renewal of growth in transport and reinforcing the growth curve in the other three sectors - this finding preceded the emergence of indicators of a "double-dip" recession in the UK.
- Specialist SRCCS were used to a far greater extent in the transport sector and increasingly in the energy and process manufacturing sectors - they featured only to a minor extent in discrete manufacturing.

- The estimated market demand for all SRCCS in 1991/1992 was of the order of £300 million (the demand for specialist SRCCS being about £75 million). This is only an order of magnitude estimate - it is reliant on estimated average system costs of some £25,000 in manufacturing and the energy sector and £100,000 in transport and an assessment of atypical sites (see the footnote to Table 4). Atypical sites in the sample appeared in all sectors, their average system costs ranging from nearly £1 million in process manufacturing to almost £7 million in the other three sectors. Clearly, relatively minor shifts in the mix of typical and atypical sites installing SRCCS can have a significant impact on the annual value of sales.

4018 The terms of reference required that any estimates relating to the SRCCS market needed to be validated. We have already indicated the difficulties likely to be associated with such validation especially in circumstances where there were doubts about the market definition. Table 5 is based on interviews with suppliers some of whom were able to provide information relating to typical system costs and market size.

4019 The evidence from the supply-side interview programme demonstrates that suppliers were more comfortable discussing highly specialised markets such as:

- road signalling systems;
- railway signalling equipment;
- gas and oil production;
- aeroengine controllers;
- naval command and control systems;
- instrumentation; and
- automotive and other engineering control systems.

Figure 3 : Structure of the Survey Sample (number of sites)

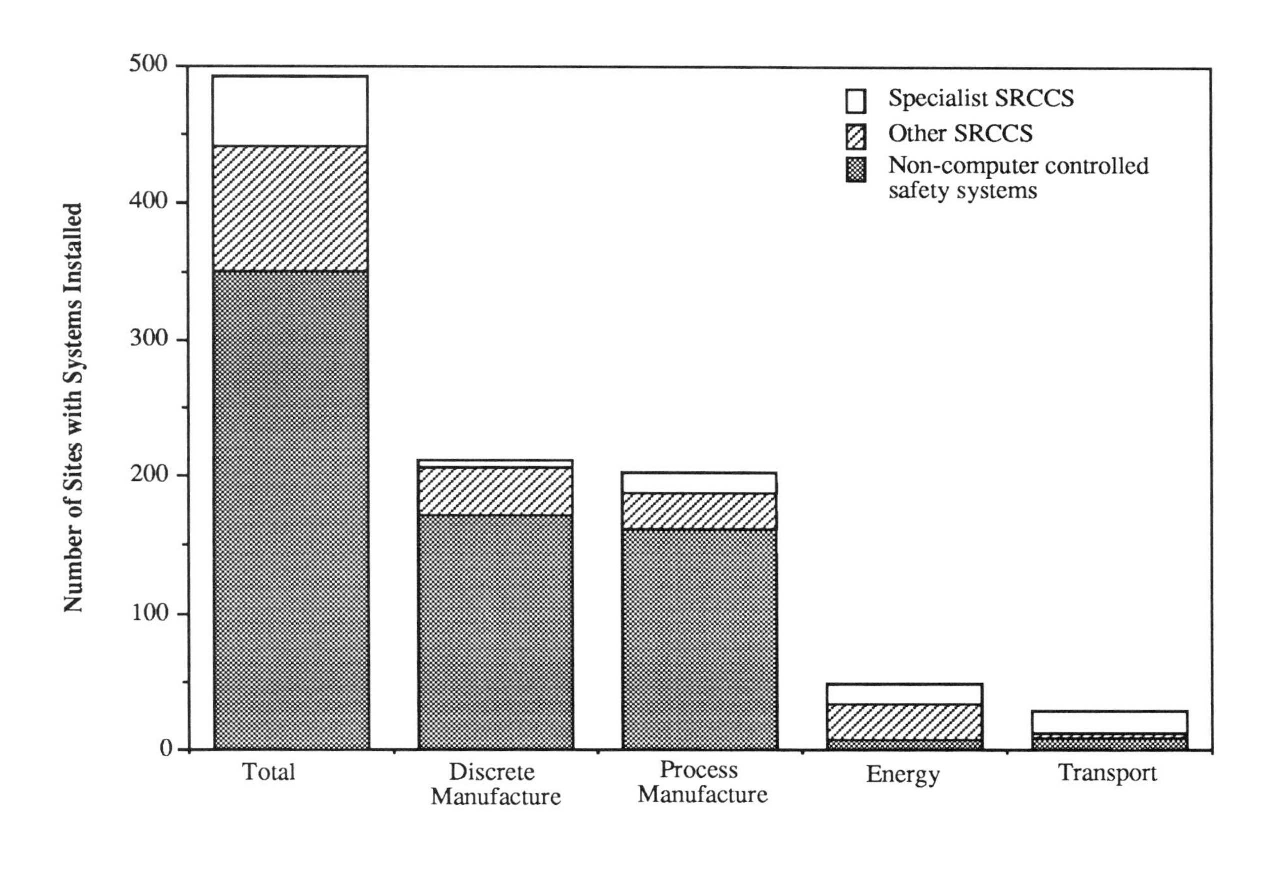

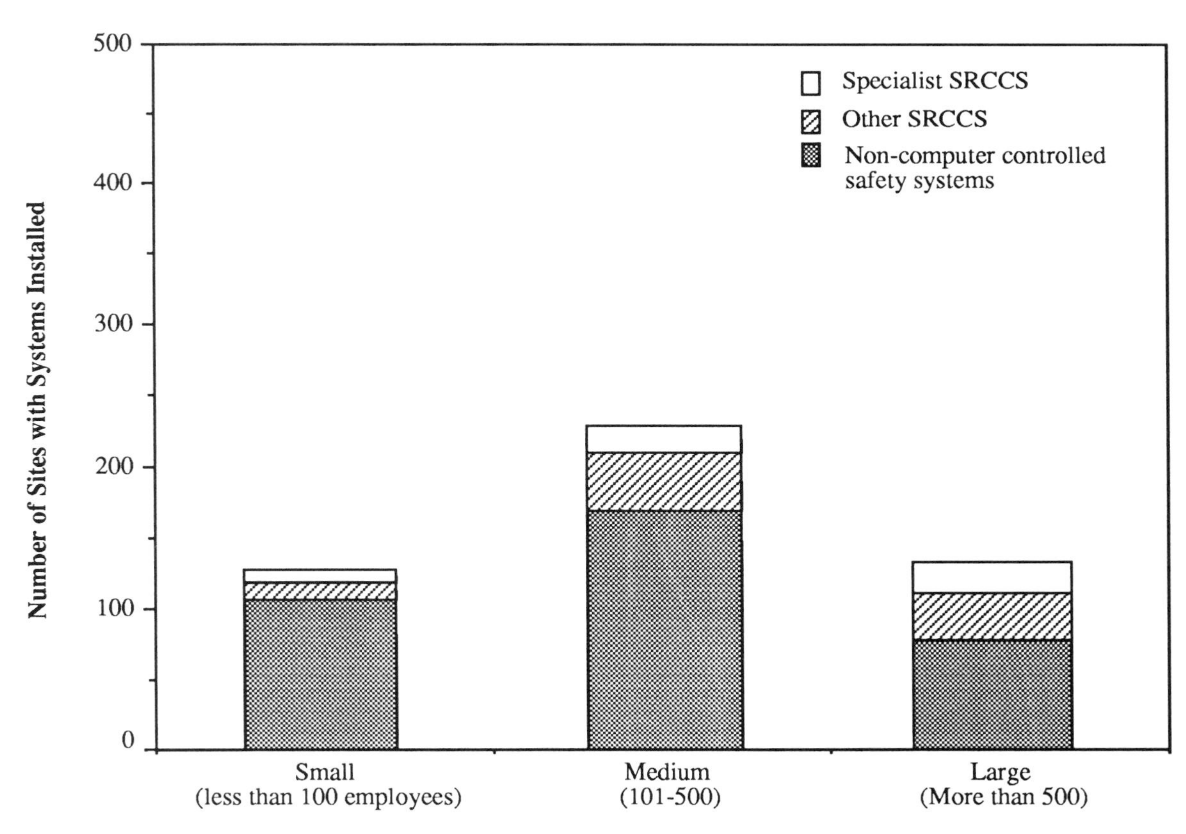

Table 3: Specific Applications of SRCCS

Sector (Number of Sites)	Applications in 10%+ of users
Discrete manufacture (39)	• Machinery (eg milling and trimming) • Iron and steel (eg casting)
Process manufacture (41)	• Machinery (eg milling etc) • Heat treatment (eg baking) • Rubber product processing • Packing/packaging/paper • Textile processing • General incl. food processing
Energy (41)	• Power generation - monitoring and control of substations • Water supply - pollution overflow • Coal mining/environment underground • Heat treatment and processing • Power generation - fuel management and production
Transport (19)	• Air traffic control • Traffic control - road and rail • Ports/shipping
Site Size (Number)	**Applications**
Small (100 or fewer employees) (16)	• Machinery • Iron and steel • Water supply
Medium (101 - 500) (60)	• Iron and steel • Rubber products • Packing/packaging/paper
Large (over 500) (55)	• Machinery • Power generation • Petrochemicals

* Sectors and site sizes in which there were 3 or more specialist SRCCS users.

Table 4: Key characteristics of SRCCS markets in the UK (1991/1992) - user analysis

	Total	Manufacturing		Energy	Transport
		Discrete	Process		
Cumulative number of SRCCS installed: • of which specialist	21,350 6,000	7,000 330	7,750 1,645	2,880 1,125	3,720 2,900
% market penetration • of which specialist systems		14 1½	16 6	83 32	65 51
Average cost per system (£000): • typical • atypical**		25 6,700	25 1,050	25 6,700	100 6,700
Market demand (£m) total: • of which specialist	300 74	130 6	80 17	50 19	40 32
Annual percentage increase in number of SRCCS installed % 1986-1991	24*	25	26	26	*

Notes:

* A high proportion of respondents in the transport sector claimed not to know when their systems were installed; those that did recorded a decline in annual numbers installed during the late 1980s compared with the early 1980s. This will have affected the total increase in numbers installed.

** An atypical system is defined to be one whose cost is significantly different from the average cost for all systems.

Figure 4 : Cumulative Number of SRCC Systems Installed; Survey Results

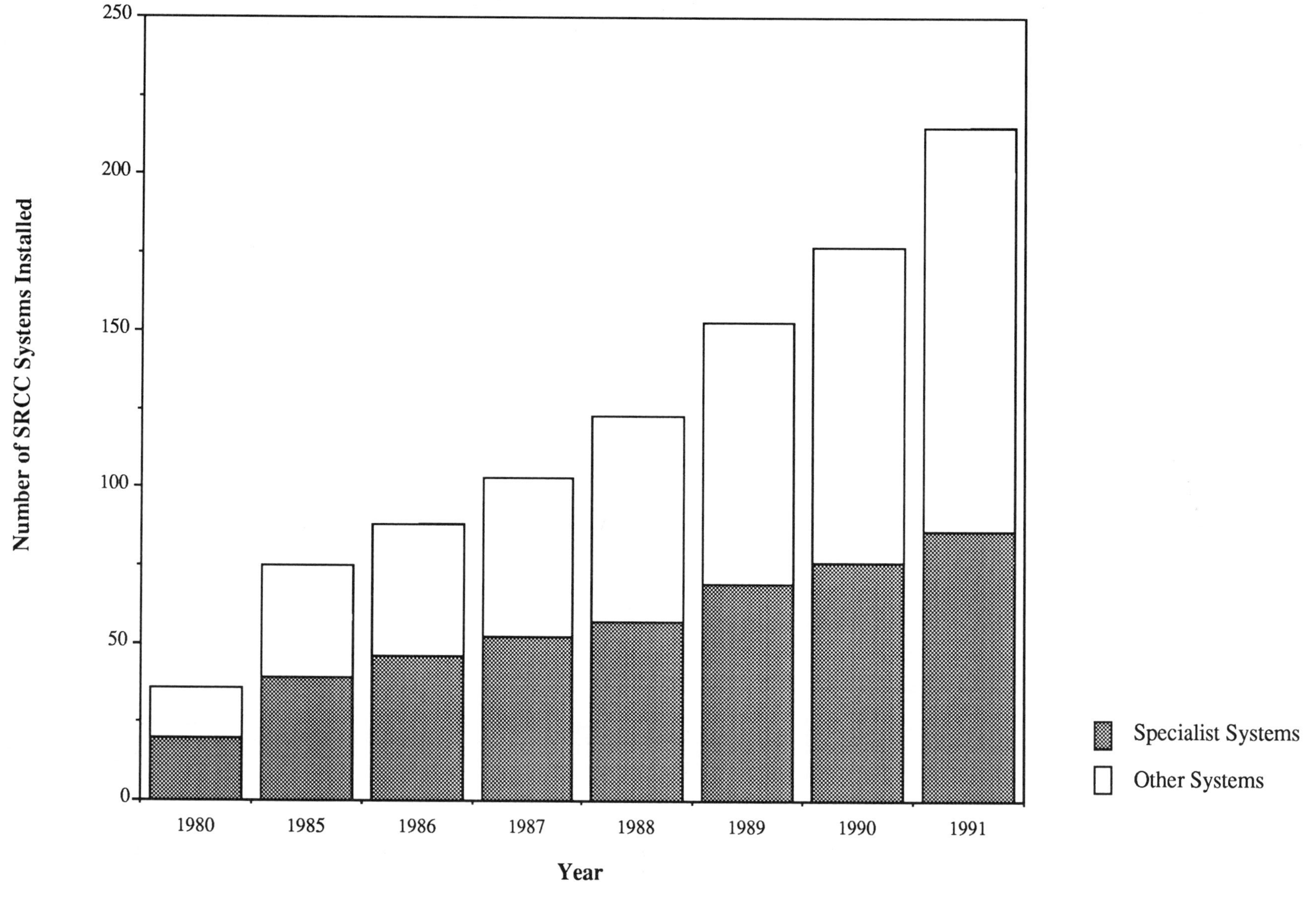

Number of SRCC Systems Installed

250

200

150

100

50

0

1980 1985 1986 1987 1988 1989 1990 1991

Year

Specialist Systems

Other Systems

GSE-32-16

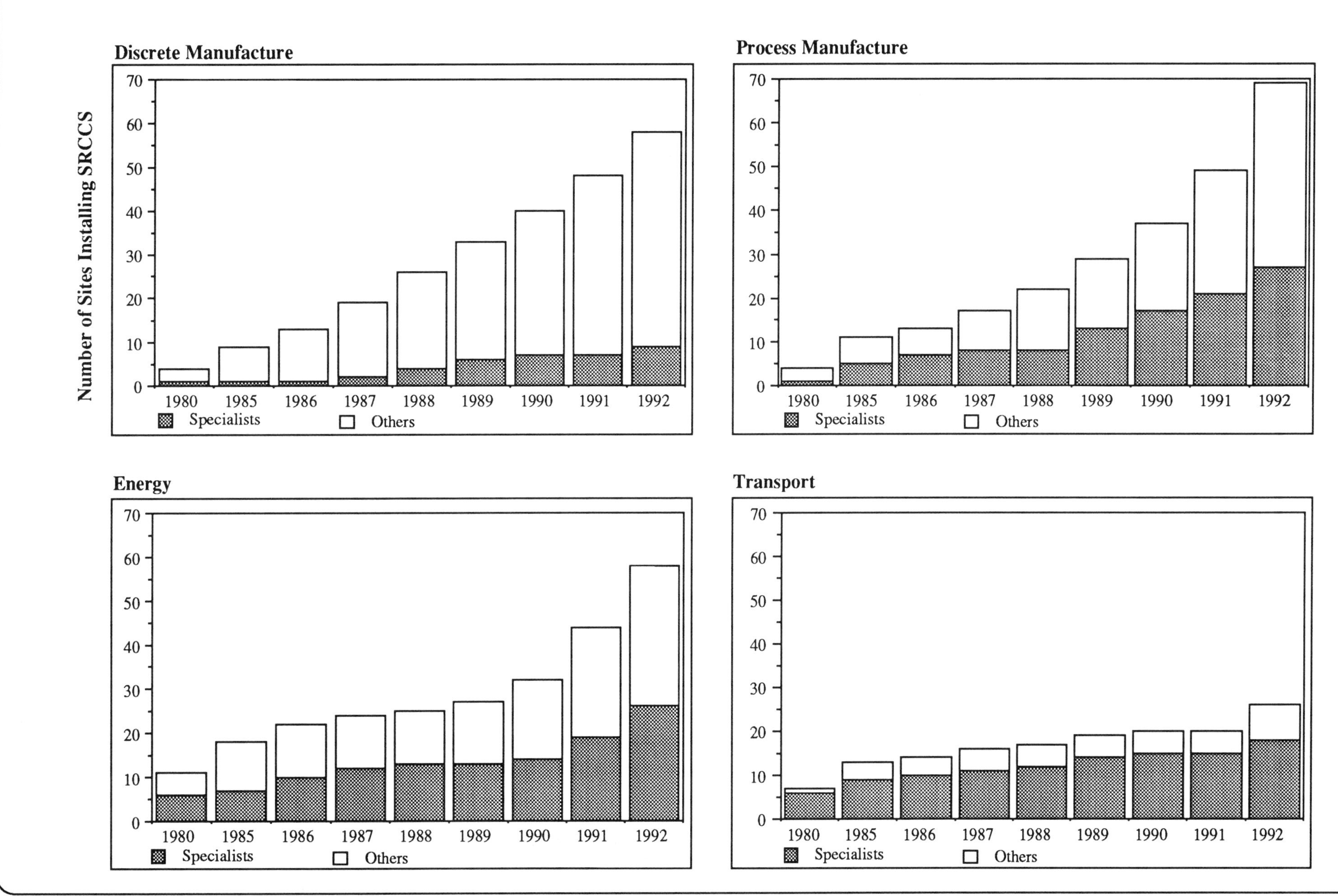
Figure 5 : Cumulative Number of Sites Installing SRCC Systems; Survey Results
Number of Sites Installing SRCCS
Discrete Manufacture
Process Manufacture
Energy
Transport
70
60
50
40
30
20
10
0
1980
1985
1986
1987
1988
1989
1990
1991
1992
Specialists
Others

GSE-32-17

Indeed, some major suppliers of control systems and software houses suggested that the SRCCS markets were too fragmented, too small and too ill-defined to warrant treatment as discrete markets. The implication was that there was more commonality between computer and non-computer safety related control systems in specific applications than there was between SRCCS in different applications.

4020 It will be apparent from Table 5 that market information from the suppliers was fragmented and incomplete. However, the following observations can be made on each of the broad sectors.

Discrete manufacturing

4021 The cost of systems ranged from £20,000 (in general engineering) to over £2 million in the defence sector. This is consistent with the demand-side estimated range. Hardware costs represented about half total system costs but software costs fell from 50% of total costs in the smaller general engineering systems to some 10% in the larger defence systems.

4022 No estimates of SRCCS demand in this sector were provided by suppliers or could be deduced. In our view this is consistent with the dominant use in this sector of non-specialist SRCCS whose safety features tend not to be marketed separately from other control purposes.

Process manufacturing

4023 The range of system costs was from £70,000 for small scale operations to £2 million for the larger ones. This was a higher range than suggested by the demand analysis. Software costs, according to one supplier, represented only 5% of the total with hardware comprising 50-60%.

Table 5: Key characteristics of the SRCCS markets in the UK; supply-side assessments

Sectors	System costs	Components of cost	UK sales
Engineering systems	£20 - 100,000	Hardware 50% Software 50%	No estimates for SRCCS separate from more general control systems
Defence systems	£2 million +	Hardware 50% Software 10% Development/ installation 40%	
Oil and petrochemicals	Small £70,000 Medium £250,000 Large £2 million	Hardware 50-60% Software 5% Other 35-45%	£40 million (6 companies' sales estimates)
Chemicals	£100,000	N/A	N/A
Power generation (nuclear)	£1-5 million	N/A	£45 million (2 companies' sales estimates)
Traffic control (road and rail)	Local £15,000 Area £50,000 - £2 million	Most hardware	£45 million (market estimate) £60 million (4 companies' sales estimates)

Notes: The assessments under traffic control do not include SRCCS for air-traffic control.

N/A: Not available.

4024 The higher average costs indicated by suppliers would suggest that our market demand figure of some £80 million could be an under-estimate perhaps by the order of £20 million. However, the sum of UK sales figures quoted by six major suppliers did not exceed £40 million.

Energy

4025 The energy sector's systems costs were based on supplier estimates from nuclear applications. These were in a narrower range than emerged from the demand analysis. The implied market demand was much the same, ie about £40-50 million. However, this could be an under-estimate because UK sales of just two suppliers comprised some £45 million (although it should be recalled that SRCCS sales reflect "lumpy" investments and that sales revenues may be scheduled over more than a year).

Transport

4026 For similar reasons as in the energy sector, the market demand figure in Table 5 could be under-estimated. However, the supply-side offered a view of the total UK market size which was only slightly higher than that suggested by the demand analysis. It is of some significance that it was only in this sector - dominated by specialist SRCCS - that a demand figure was provided by suppliers. It must be emphasised that neither the demand nor supply side estimates adequately reflect sales of systems for air-traffic control.

Market demand and supply in the UK and elsewhere in Europe

4027 If the demand analysis estimates of market size are adjusted by the views from the supply side, the annual demand for all SRCCS in the UK could be some £350 million, of which about one quarter could be represented by demand for specialist systems.

4028 Information for a robust analysis of the supply chain used to meet this demand in the UK was not available either from the demand or supply side of any of the SRCCS markets. Indeed, suppliers who might have been expected to be a valuable source of evidence on supply chains were rather vague in their responses on this, especially with regard to the component, software and development inputs required. Thus, the evidence available did not permit an investigation of the supply lines prior to system integration or manufacture.

4029 In one sense this is not surprising. What the study found is that the fragmentation of SRCCS markets is compounded by the diversity of players in the various markets. These players will include the control system suppliers (few of whom will straddle more than one SRCCS market), software houses, engineering contractors, the hazard and safety assessment consultants, the training providers and research and development organisations. Each SRCCS market is likely to be characterised by a particular confluence of all these players, few of whom will be dedicated to SRCCS.

4030 Given the characteristics of the market as described above and the poor response to our non-UK European survey, it has not proved possible to provide independent estimates of the market for SRCCS in Europe. Our inquiries of trade associations in Europe suggest that the application sectors for SRCCS may not be much different from those in the UK - very high penetration in the energy and petro-chemical sectors, high penetration in transport and low penetration in discrete manufacturing such as defence and automotive products. If the same market segmentation exists elsewhere in Europe as in the UK there will be similar difficulties in estimating market size in terms of the number of installed systems. Moreover, any such market estimates are likely to be of limited current relevance. The niche SRCCS markets are dependent on close supplier - customer relations and do not lend themselves to much cross-border trading. It is worth emphasising that few on the UK based supply-side were anticipating exports elsewhere in Europe and few amongst the users had relied on imported systems.

4031 One estimate of the European market for process control equipment quoted to us during the supply side investigation was £550 million in 1985 of which £200 million could have had a safety related component. If the market has been growing at the same rate as estimated for the UK (ie some 25% per annum over the last half of the 1980s - see Table 4), then this might suggest a European market of nearly £1,000 million in current prices. If this is equivalent to the estimate for the UK market, it implies a UK market share in Europe of 35%. Although it was claimed by some firms that the UK was well in advance of other European countries in the off-shore market for SRCCS and perhaps also in petrochemicals, this market share estimate seems high compared, for example, with estimates of a 20% UK share of Western European markets for control and instrumentation in the late 1980s.

Hazard analysis and risk assessment

4032 Hazard analysis and risk assessment are fundamental to the system safety approach. Together they determine the safety integrity requirements of systems. In turn, this drives engineering solutions to assure safety.

4033 From our supply side assessment, it would appear that only six amongst the UK organisations interviewed claimed to be involved in undertaking formal hazard analysis and risk assessments for users. Of these, four were consultancy houses or other influencers that undertake independent third party assessments. However, this type of activity only accounted for a small proportion of their total business. This limited demand was reflected in the responses to the demand side survey:

- very few users (mostly in process and energy sectors) undertook hazard analysis and risk assessments with most relying on their own past experience in using systems to deal with potential hazards; and

- those who undertook hazard analysis and risk assessments tended to use in-house resources and expertise.

4034 The requirements specification was seen by suppliers as the weakest link in the life cycle chain and the most fundamental. This was confirmed by information gathered in the user survey. The results suggested a degree of complacency with respect to risk. There was little evidence of any explicit and systematic application of risk analysis techniques by users. This appears to highlight a significant training and education issue.

4035 The specifier of the type of system to be installed was supplied from in-house expertise in the overwhelming majority of SRCCS users in the survey. Only one third used specific techniques to specify their requirements - 40% in the case of specialist SRCCS users. Formal hazard analysis had been used on only two sites - both in the energy sector. A large minority (two-fifths) of those claiming to use specific techniques could or would not say what they were.

4036 Users' evaluation of system requirements in terms of the severity and frequency of potential hazards on this evidence appeared to be undertaken in a rather subjective way, expressed in qualitative rather than quantitative terms.

4037 Users in the survey were able to provide a picture of their existing systems in terms of the 'frequency probability' and nature of the hazards they were intended to protect against (see Table 6 where these terms are defined). This gave an indication of the type and criticality of application of these systems by sector (see Tables 7 and 8). In broad terms, the tables show that SRCCS were used:

- in discrete manufacturing to protect operators against fatal injury across the range of frequency probabilities of system malfunction;
- in process manufacturing to protect against a low frequency probability of hazard involving potential fatal injury to the operators and the general workforce as well as damage to plant;
- in the energy sector to protect against a wide range of frequency probabilities with some emphasis on high frequency probability of process malfunction, plant and environmental damage; and
- in the transport sector to protect the general public against injury arising from a high frequency probability of process malfunction.

Table 6: Number of SRCCS in each criticality class: user survey results

Potential Hazard	Frequency Probability					Total
	Frequent	Probable	Occasional	Remote	Improbable	
Process malfunction	38	12	16	26	8	107
Plant damage	32	13	14	32	7	102
Operator fatal injury	31	9	16	26	5	94
Non-operator fatality	34	6	8	19	3	74
Workforce general injury	38	8	13	19	7	86
Environmental damage	16	7	8	24	6	65
Spillage of dangerous substance	9	9	7	22	5	55
General public injury	19	1	2	10	5	39
Total	46	14	20	39	11	141

Note: Frequent - many times a year

Probable - at least once a year

Occasional - once during product life-time

Remote - unlikely

Improbable - incredible

Table 7: Frequency probability (% users)

Frequency	Total	Discrete Manufacturing	Process	Energy	Transport
• At least once a year	47 (58)	34 (25)	32 (40)	56 (56)	89 (87)
• Occasional	14 (10)	26 (25)	12 (13)	12 (13)	- (-)
• Remote incredible	38 (34)	36 (50)	52 (53)	39 (38)	5 (7)
• Don't know	6 (6)	5 (-)	10 (7)	5 (6)	5 (7)

Note: % of users will not sum to 100 because each site may have more than one system covering different frequencies of potential hazard.

Figures in brackets refer to the % of specialist systems.

Table 8: Nature of hazard potential (% of users)

	Total	Discrete Manufacture	Process	Energy	Transport
Process malfunction	76 (82)	64	63 (67)	90 (94)	95 (100)
Plant damage	72 (72)	64	76 (87)	88 (81)	47 (47)
Operator fatal injury	67 (58)	90	85 (93)	39 (38)	42 (40)
Workforce fatal injury	61 (52)	59	73 (67)	56 (44)	47 (40)
Non operator fatal injury	52 (50)	59	61 (60)	39 (31)	53 (53)
Environmental damage	46 (48)	26	46 (60)	78 (81)	16 (7)
Spillage of dangerous substances	39 (36)	31	41 (53)	61 (56)	- (-)
General public injury	28 (38)	13	15 (13)	27 (25)	89 (87)

Note: Figures in brackets refer to specialist system users - no figures for discrete manufacturing are given because of small sample size.

4038 Amongst the users in the sample, the decision to opt for an SRCCS solution to the identified requirements was driven primarily by the imperative to minimise the risks and comply with the regulations associated with the nature of the process or the degree of hazard. Cost-benefit analysis was a second-order factor in the decision. A factor of particular importance in manufacturing (discrete and process) was the facility offered by SRCCS of integration into total system control. This supports the view that demand for non-specialist systems does not represent a specific safety-related market.

4039 Although this evidence - combined with that on specification techniques - does not suggest that hazard assessment was generally used in specification of requirements and solutions, nevertheless training in hazard assessment was provided by some 70% of users of SRCCS - 85% in the transport sector. Training in system specification, on the other hand, was provided by a much lower proportion of sites - about 40%. These figures were no higher for the sites using specialist SRCCS. The extent of reliance on in-house expertise for specification of the systems would not appear to have been supported by widespread training in system specification or use of hazard assessment techniques even for specialist systems.

Strategies for risk reduction

4040 The evidence on use of hazard analysis and risk assessment of system requirements, raises some concern about the ability of users to identify SRCCS as the most appropriate risk reduction strategy and subsequently to specify safety design and integrity requirements correctly.

4041 Users in the survey did experience design difficulties and a lack of necessary in-house design skills to support the specification process but only some 20% of the sample claimed this - a higher proportion (about one-third) seeing a lack of in-house skills as a particular problem for dedicated systems. Generally, users were sanguine about problems posed by the specification of systems. Nearly 30% claimed to have experienced no problems.

4042 By contrast, of those that provided information from the supply side, 22 respondents (nearly 75% of which were systems manufacturers or engineering contractors) expressed considerable concern that the specifications supplied by users were ambiguous and generally of very poor quality. This was attributed to a lack of user expertise. Consequently, considerable effort was required to translate the specifications into system engineering requirements for hardware and software. Although suppliers considered that they had reasonably close relationships with their customers, there was still seen to be a real danger arising from engineering specifications that were incorrect. Their view was that more attention needed to be paid to the specification stage and that users should adopt more rigorous and comprehensive system design and specification approaches. Clearly, close supplier/user interface is critically important at this stage.

4043 While a number of tools and techniques are available to assist in the development of usable specifications, only a third of the supply side respondents, excluding the regulatory authorities, gave any indication that they were used. All the suppliers were asked about techniques used and most of their responses were generic rather than specific. The specific techniques most often quoted as being used were the first three in Table 9.

Table 9: Specification development techniques used: supply-side assessment

Yourdon design methodology
In-house tools
Z formal specification language
CAD design tools
Fault tree analysis
HOL
Lifespan
PCMS
SPARK
SSADM
Teamwork

4044 Use of these techniques allowed the supplier and the user to examine the original specifications for fundamental omissions and misconceptions. Prior to any subsequent system development work the revised specifications would be agreed between the user and the supplier.

4045 In undertaking development work to meet specification requirements, suppliers were aware that the appropriate strategy to reduce risk could be to use non-computer based technology eg protection systems based on physical barriers, hardwired interlocks and mechanical reliefs. Both the users and the suppliers suggested a large part of the reason for this was that the adoption of a non-computer controlled strategy was sufficient to meet requirements, given the frequency of risks and hazards involved and the cost/benefit implications of using SRCCS. Three particular factors tended to favour the use of non-computer based protection systems:

- there was some anxiety among users that computer controlled systems needed to be fail-safe but might not be able to guarantee that degree of reliability;
- it was easier to demonstrate, and provide assurance of, the reliability and dependability of non-computer controlled systems; and
- the capital costs were lower for non-computer controlled systems.

4046 These particular concerns were alluded to at the focus groups. One respondent currently using a non-computer controlled safety system suggested that purely on safety grounds his company saw little reason to change to a SRCCS:

> "I've got systems that I trust. Never failed me in the past. Can't see them failing me in the future - why should I use an electronically based system?"

4047 Moreover, anxiety was often expressed about allowing a SRCCS to operate in a safety function because it was not possible to ensure that the software was as reliable and dependable as the existing non-computer based safety system. The result, according to a respondent from the manufacturing sector, was that:

> "I cannot see, certainly in our work situation, where anybody would release a computer to actually take over control (of safety)."

4048 In the light of these observations, it is therefore not surprising that many of the systems manufacturers and integrators supply hardwired safety-systems as well as computer based safety systems.

4049 Suppliers also suggested that in their view many companies only considered the capital costs associated with the strategic options that could be used to reduce risk. They felt that a full cost/benefit analysis was not usually undertaken. As a result, little or no consideration was given to assessing the options in terms of the additional costs of:

- full life maintenance;
- installation;
- operator and maintenance training; and
- lost production.

Suppliers suggested that if all the costs involved were considered, then the benefits of using safety-related computer control systems became more evident. Users that had undertaken some form of cost-benefit assessment confirmed this during the focus group meetings, clearly indicating that use of computer control for safety has yielded, in some cases, considerable added value. In particular, computer control systems were seen as significantly increasing productivity, particularly as far as maintenance is concerned. For example, the replacement of a mechanical control system with computer control can reduce the frequency with which it is necessary to undertake routine plant maintenance.

4050 It should also be noted that, not only was there limited use of a full cost-benefit appraisal, there was also a significant degree of ignorance about the options. Only about 30% of the non SRCCS safety system users in the survey were aware of any SRCCS system. Of those who were aware of SRCCS, only one quarter had considered adopting them. Non-users in general, according to the survey, were unlikely to be a source of increased demand for SRCCS over the short to medium term.

4051 The final stage in the risk reduction strategy is the decision to procure and install a specific SRCCS. Depending on the exact application requirements, users procured safety-related computer controlled systems via one or more of the following routes:

- their own internal development department(s);
- directly from systems manufacturers, system integrators or software houses; and
- indirectly through third party engineering contractors.

4052 In the latter two cases, the suppliers, especially the systems manufacturers and engineering contractors, indicated that users tended to have an approved vendor list of potential suppliers. This was also commented on at the focus groups. This will comprise those suppliers:

- which users had sourced from on previous occasions;
- whose systems and hardware and software components had been tested by users and had been "approved" for use;
- who could demonstrate that their systems were designed, developed and produced so that they:
    - were easy to operate and maintain;
    - had a high level of flexibility (ie capable of being adapted to changing operating environments);
    - were able to meet user requirements; and
    - had a high level of reliability.

4053 The user survey indicated that ease of installation, price and cost-effectiveness were the least important factors which determined the choice of SRCCS and of the supplier. Given the relative importance of non-price factors and the rapid developments in computer technology, users would be at a disadvantage in the procurement of systems if they lack information and awareness of SRCCS. They were reliant on the correct specification of system requirements and on the suppliers' capabilities to interpret and subsequently design, develop and install systems that met the safety requirements.

4054 Suppliers, particularly the systems manufacturers, indicated that greater emphasis was being placed on third party approval and certification mechanisms to demonstrate more clearly their quality of approach. Obtaining certification to ISO 9001 (BS5750 Part 1) or AQAP standards and using DEFSTAN 00-55 were seen as ways of achieving this.

4055 While certification or use of these standards acted as a useful signpost for users, there was still concern over the lack of information about SRCCS and the issues raised. Some 40% of SRCCS users indicated that information was difficult to find and that more information was required. The main requirement was seen to be to obtain general information on everything likely to be of relevance. The relatively high demand for more information of a general kind was indicative of a lack of awareness of the issues raised by SRCCS and was reinforced by the proportion of users and non-users expressing an interest in joining a safety-related computer controlled systems association or club - 20% amongst non-users and nearly half of the users. In contrast, only two supply side organisations gave any indication that they faced a similar information shortfall and that more information was required.

4056 The association or club referred to in general terms in the market survey has of course since been established. It is called the Safety Critical Systems Club which is sponsored by the DTI and the Science and Engineering Research Council under the Joint Framework for Information Technology.

4057 The evidence of this study is that strategies for risk reduction using SRCCS were likely to have been inadequately thought through because:

- there was a significant degree of ignorance about the strategic SRCCS options amongst those who operated safety systems;
- hazard assessment and cost-benefit analysis were not used to any great extent in system specification and design;
- although there was a general tendency to use in-house resources for system specification, design and validation, there was limited training provision on these matters.

4058 There is no evidence of these factors having an adverse impact on the performance of SRCCS. The point being made here is that users are not generally making fully informed choices about whether to use SRCCS as part of their risk reduction strategy.

Tools and techniques for safe systems

4059 From the desk research for this study, it would seem that the main focus of work to date in SRCCS has been technological to develop tools, techniques and procedures designed to assure the safety of the systems. This is demonstrated most clearly through the proliferation of various guidelines and standards encouraging the use of various tools, techniques and procedures in the design, development and installation of systems. A detailed review of existing and emerging standards relevant to SRCCS is presented in Appendix G. It is worth emphasising that only 20% of the survey sample of safety system users (including non-SRCCS) were aware of SRCCS related standards and only half of the SRCCS users was aware of such standards. Of those that claimed awareness of the standards, nearly one third cited the HSE 1974 regulations and about one fifth quoted non-specific UK standards. So, according to the survey there was little awareness amongst users of the specific standards reviewed in Appendix G.

4060 For organisations on the supply side, particularly the systems manufacturers, integrators and software houses, the various guidelines and standards now have to be put alongside those developed in-house. Where information was provided all the organisations approached on the supply side indicated that they already work to in-house guideline and standards and that these had been and were being supplemented in the light of external developments. In particular, suppliers were introducing, and placing increased emphasis on, quality assurance procedures, the most formal manifestation of which was accreditation to BS5750 or ISO 9000. A quality assurance approach facilitates traceability when failures occur. But suppliers recognised that attention must also be given to the design approach and the way in which the systems could be proven to be safe.

4061 It was emphasised that adherence to standards did not necessarily result in the production of a safe system. The approaches used in the various standards developed to date were regarded as procedural; they focused on the process rather than the goal of a safe system. As such, while this encouraged a more systematic approach to the design, development and implementation of systems it did not guarantee acceptable safety levels.

4062 Some of the standards were regarded as being more prescriptive in their approach than others in terms of specific tools and techniques that should be used to supplement the process by which systems evolved. A particular example of this usually referred to by suppliers was DEFSTAN 00-55, which was seen as emphasising the use of formal methods. While there were mixed feelings about the practicality of adhering to DEFSTAN 00-55, its more "goal oriented" approach was considered by many to be more useful than simply specifying the methodology.

4063 There was a general view expressed among suppliers, although not particularly strongly, that for SRCCS applications high-level modular software languages should be used as a means of reducing, or being able to detect more readily errors likely to occur in the development of software code. This was supplemented by reusing code known to work in other similar applications. In the majority of cases, however, respondents provided little indication of the software languages used in practice. For the small number of cases (ten) where information was provided, Ada was the preferred language (used by seven of the ten). For most applications, there was little or no indication that there was any particular external pressure to use Ada. However, in military applications use of Ada was seen as being almost mandatory.

4064 Among systems manufacturers, reference was sometimes made to the use of "standard industry software" which could be used "off the shelf". In such cases there was little indication that it mattered what software language was being used, as long as it could be used, had a proven operational track record, and was acceptable to the client.

4065 Relatively more attention was given to the way in which the software programs were verified and validated. A variety of tools and techniques were used for this and there was little to indicate whether one approach dominated. Reference was made to the use of CASE tools in a generic sense, and a limited number of respondents did indicate the use of Lifespan, OBJ and Yourdon and the static analysis tools MALPAS, SPADE and Logiscope. In addition, use was also made of in-house tools which had been developed.

4066 In the context of the verification of software programs in particular the software houses referred to the potential of formal methods. In many cases the main pressure to use formal methods was coming from the defence sector following publication of DEFSTAN 00-55. More generally, formal methods were seen as one way of providing greater assurance that errors present in SRCCS had been minimised both in terms of both the design of the software and the hardware.

4067 In practice few suppliers, particularly outside the software houses and consultancies, used a formal methods tool. To a large extent this was because formal methods were relatively new and the individuals involved in the verification and validation of systems had not been exposed to these tools before. In addition, the cost of training associated with the use of formal methods was regarded as prohibitive.

4068 The other key component in the systems manufacturers' and integrators' approach to developing safe systems was seen in their approach to their use of hardware and its configuration and architectural designs. This was less of a concern to the software houses than systems suppliers and integrators, except where they undertook third party verification and validation. The overall objective in this area was to produce hardware that was highly reliable and dependable. The principal means of achieving this were to:

- use standard field proven or "approved" hardware components; and
- design and build systems to be fault tolerant through the application of parallel microprocessors.

4069 A further stage of testing was undertaken at the systems integration stage of the system lifecycle ie factory tests. Standard in-house test procedures were used. They included systems simulation; heat soak, vibration and temperature tests. If required, tests specified by users would also be undertaken. However, little information was given to indicate the extent of the testing or how reliable it was.

4070 As an extension to in-house verification, validation and testing, suppliers also noted the attraction of having systems tested and approved by an independent third party. Several suppliers had applied for, or were in the process of or were actively considering applying for German TÜV approval for their systems. These suppliers believed this was the best route for proving that the systems they produced were safe. Suppliers gave no indication of there being any generally recognised alternative approval body to TÜV, either in the UK or elsewhere. A review of the German TÜV systems for accreditation of SRCCS is provided in Appendix H. The suppliers did not indicate what aspects of their systems were being sent to the TÜV for approval but they were not concentrated in any one particular sector.

4071 Installation of the final system was typically undertaken by suppliers either directly using their own engineering staff or indirectly using engineering contractors. Where engineering contractors were used, they would be supervised and inspected by the supplier. According to the suppliers, it was unusual for the user to install the system and, where it happened, the work would be supervised and inspected by the supplier. The survey suggested that user installation was more common than implied by the suppliers - for nearly 20% of users (nearly 30% in process manufacturing) the system installer was in-house.

4072 Both suppliers and users were involved in the final validation of the system. This typically involved a series of system and component simulation tests. Users, particularly in the energy sector, also undertook final instrumentation quality checks and tests using specific in-house guidelines. While these tests were carried out, suppliers provided a hotline service for trouble-shooting support. However, as with the factory tests, there was no clear indication of how extensive or effective these tests were at providing assurance that the system was safe.

4073 In general, the supply side respondents gave the impression that those involved in the design, development and production of SRCCS were competent to do so. They were of the opinion that to produce safe systems required individuals who had a sound educational background from a narrow range of engineering disciplines supported by relevant experience. This was reflected in the educational qualifications and experience of those employed to work in the SRCCS field. Of the 64 respondents, 30 indicated that the basic qualification requirement was to be an engineer. A further 23 of the respondents were more explicit, indicating that chartered status or more generally, a graduate level engineering qualification was needed. The engineering disciplines identified by respondents as being necessary to work in the safety related area are presented in Table 10. It reveals a distinct preference for computing and computer science engineers and electronics engineers.

Table 10: Skills background: the SRCCS supply-side

Appropriate Skills	Number of References from Supply-Side Respondents
Computing/computer science	12
Electronics engineering	11
Software engineering	5
Electrical engineering	4
Mathematics	2
Mechanical engineering	2
Others of which:	
- instrumentation engineering	1
- control engineering	1
- rail engineering	1
- chemical engineering	1
- physics	1

4074 In most cases, the emphasis was on being able to integrate individuals who had appropriate practical engineering skills, rather than academic or scientific knowledge, into a team. This was considered as necessary because SRCCS related work covered a range of disciplines. However, these skills had to be supplemented by "relevant" experience. What was meant by relevant experience was not always made clear, although in some cases the suggestion was that the individuals concerned should have had three or more years applying their skills in an industrial environment.

4075 Without exception, further training of recruits was seen by suppliers as essential. This was provided through on-the-job instruction, supplemented by external sources as appropriate. Respondents were, however, rather vague as to what were their training requirements, what training was provided and how extensive the training should be. Those providing some indication of the type of training undertaken to date made reference to training in:

- the use of different software languages;
- the use of design development tools;
- quality assurance procedures;

- relevant standards;
- personal skills development (eg presentation and writing skills); and
- the use of the hardware, or in some cases the software, supplied by third party vendors (this particularly applies to the use of microprocessors and process logic controllers).

4076 There was no indication that software engineering was a cause for concern despite close questioning of the issue. On the other hand, some of the software houses raised the more general concern that the software industry was very immature and had not yet developed a structured professional approach to the development of software programs. This weakness was compounded by the fact that anyone could write software; no form of qualification was required, a situation which many of those directly involved in the development of software described as bizarre. Where this issue was raised, training was rarely indicated as a solution to this situation. Those that did make suggestions were of the opinion that software programmers should only be able to work on safety-related projects if they could demonstrate they were competent to do so. This required the development of a testing and certification process - not training per se.

4077 Certification was seen as one of the best ways of ensuring that software would be written correctly and help ensure safety. It would also ensure that verification and validation procedures would be more closely followed. There could also be some benefit as a marketing tool; having formally certificated staff could act as a flag for users attempting to distinguish between suppliers. An alternative or supplement would be for suppliers to obtain accreditation to BS5750 pt 1 and TickIT accreditation for software.

4078 This is not to say that training was regarded as unnecessary. The majority of systems suppliers and software houses claimed that more could be done by educational establishments by providing wider courses in subjects such as modelling, data structuring, design methods and the ability to work effectively as a team. This contrasted with the views expressed by the training providers and academic institutions who felt that courses were now better geared to meeting these sorts of requirements. The difference in views may however be explained by the timelag involved in filtering the benefits of improved and expanded courses to industry. A change in course content will not be visible to industry for three to five years, or more, depending on how long it takes for individuals to complete courses and become employed.

4079 More specifically, the systems suppliers, and the software houses in particular, took the view that safety-related issues (eg reliability and risk assessment) were not integrated into existing engineering, computing, electronics and software engineering courses. This was confirmed by other evidence. However, it was pointed out by one respondent that this approach may be of limited value because safety cannot be considered regardless of the context in which it is required; safety in practice is generally application specific.

4080 The evidence from the user survey is that there is not a strong demand for safety specific training. Only about one half of SRCCS users provided training in system installation and validation and less than one half (but more than one third) in system specification and design. This suggests that the extent of reliance on in-house expertise for system specification may not be supported by adequate training.

4081 This was confirmed by the low proportion of users (17%) who relied on external sources of training alone and by the high proportion (some 55%) who expressed no enthusiasm for more formal training (about 30% of whom said that "safety was already built into the system").

4082 On the assessment of this section any problems with regard to the effective application of tools and techniques for safe systems are likely to stem from:

- a growing proliferation of guidelines and standards with which users are often not familiar;
- uncertainty about where the responsibility for safety lies;
- the relative infancy of the software industry and the lack of a certification process for individuals or suppliers; and
- a low level of demand for safety specific training, particularly by users from external sources.

Safety in operation and maintenance

4083 The final phase of the systems lifecycle involves operation and maintenance of systems. This phase is one in which suppliers felt themselves to be vulnerable. Whilst they may have produced and helped to instal systems that are safe, any improper use or maintenance could have potentially disastrous consequences. Some of the systems suppliers considered that, if accidents occurred, most would be as a result of unsafe operation or maintenance which changed some parameter of the system without assessing the implications of that change. Indeed it was only suggested once during the supply side interviews that the use of an unsafe SRCCS was to blame for the occurrence of a hazard.

4084 Despite this concern, suppliers gave very little indication that they had made users aware of the situation. Much of the training provided by systems suppliers for users concentrated on the functionality, operation and performance of the systems. More technical training for user maintenance personnel was also provided and system operator and maintenance manuals were made available. Training was usually tailored to the individual user and could be on or off site as required.

4085 Little indication was given as to the depth of the training provided or whether it was considered effective. Where specified, the usual length of an operator or maintenance training course given was three to five days. The courses appear to be deliberately kept short because suppliers' experience suggests users regard anything longer as too expensive in terms of "lost time".

4086 Training provided by software houses to clients was said to be intensive. It was usually application and project specific and would seek to ensure that client staff were capable of using the particular software language and method used and that they recognised and understood its design. It was noted by one software house that the extent to which client training was provided was sometimes limited because budget restrictions tended to bite on training before anything else.

4087 Nearly 90% of users in the survey provided training in system use and nearly 80% in system maintenance. However, as already noted little reliance was placed on external sources of training alone (only 17% of users). We suspect that this reflects the tendency for users to send a small number of operators and maintenance supervisors and managers on supplier training courses who then pass on their acquired knowledge in-house. The relatively low level of commitment in terms of person days (35% of users devoted only 1-15 days to training) and investment in training (13% invested less than £15,000) by users tends to support this view.

4088 Overall, users considered that the training provided in the use and maintenance of systems was very useful. However, a majority of users expressed little enthusiasm for the provision of more formal training, other than to keep up with technical advances. Part of the reason underlying this lack of enthusiasm for more formal training was the belief that:

- the training currently provided was sufficient;
- safety was already built into the system; and
- the danger already inherent in the industry had already established the need for training.

4089 These responses suggest that users could be complacent about the training necessary for the safe operation and maintenance of safety-related computer controlled systems. However, it was pointed out by a number of organisations on the supply side that any apparent lack of enthusiasm for training at present was likely, at least in part, to be a reflection of the current state of the economic environment.

External factors and their impact

4090 In the preceding parts of this section the size and structure of the markets for SRCCS have been examined along with aspects of their internal operations such as the use of training. The picture that has emerged of the SRCCS markets is one characterised by:

- no well-defined single market for SRCCS but rather, first, a market for computer control systems blurred at its boundaries where safety related features are one amongst many other features and, second, a set of discrete markets (niche sectoral markets) for specialist systems where safety is the only or primary purpose;

- the more general safety related market having an annual turnover in the UK in the early 1990s of some £260 million and the set of specialist markets of about £90 million (significant caveats must accompany these estimates) and the overall SRCCS market growing at about 25 % per annum in the late 1980s;

- lack of awareness amongst potential users of both SRCCS systems and standards such that little demand for SRCCS is expected from current non-users;

- the fragmented, self-enclosed nature of the SRCCS markets, especially for specialist systems, leading to limited marketing efforts by suppliers outside their existing customer base and to a "do-it-yourself" culture amongst suppliers and users with regard to system specification and training; and consequently

- an increasing proliferation of guidelines and standards with regard to system design tools and techniques and limited demand for and supply of external training.

4091 In the final part of this section the external factors are considered which are likely to exercise a significant influence on the SRCCS markets as described above over the next decade. There are three broad categories of such factors:

(a) legislative pressures and the role of particular regulatory authorities;

(b) the development and implementation of guidelines and standards; and

(c) the completion of the Single European Market.

(a) Legislative pressures

4092 A number of supply side respondents were of the opinion that legislative pressure was likely to drive industry towards improving the level of safety. In some sectors such pressure was particularly strong because of the role and power of a regulatory authority. This was said to be the situation in the following sectors:

- nuclear;
- offshore oil and gas;
- civil aviation; and
- rail.

4093 In each of these sectors, the relevant regulatory authority (Nuclear Installations Inspectorate, Offshore Inspectorate, Civil Aviation Authority and Railways Inspectorate) were involved, to a lesser or greater extent, in providing approval for the use of SRCCS. As such, they were usually involved in providing commentary or advice on the specification and design of such systems.

4094 Industry more generally was influenced by the role and activity of the Health and Safety Executive (HSE) through the implementation of the Health and Safety at Work Act 1974. The more specific influence of the HSE in terms of SRCCS was felt through the publication of guidelines for programmable electronic systems. Suppliers showed a high level of awareness of these guidelines. However, because it has no statutory power to enforce the use of these guidelines, the HSE's influence was not considered to be particularly strong.

4095 The impact of these legislative pressures as currently perceived is likely to reinforce the self-enclosed nature of the specialist markets as they become manifest in application specific standards and guidelines. Any broadening of the SRCCS market basis, breaking down existing market barriers, will depend on the extent to which the different standards can learn from each other and on the degree to which the HSE helps this process by active promulgation of its guidelines and by the development and implementation of a training and certification strategy.

(b) Guidelines and standards

4096 Awareness of standards and guidelines amongst users was generally low even for those standards and guidelines which were specific to the energy and transport sectors (see paragraph 4059). What was even less clear was the depth of knowledge about them. For example, a number of respondents indicated that DEFSTAN 00-55 was simply about formal methods; it is not.

4097 The evolution of sector specific standards and guidelines was considered to present minor problems for several systems suppliers. It was considered that, for systems capable of being used in applications in different sectors, the suppliers will either need to:

- produce the system using different design, implementation, integration and testing procedures; or
- develop a system capable of meeting two or more existing standards.

4098 There appeared to be some consensus among supply side respondents that, because of the variety of applications of safety related computer controlled systems across different industries, a generic standard for the development of safe systems was probably necessary. Little or no indication was given of the best approach but industry consultation was claimed to be necessary. This particular consideration was of significance for respondents because some standards were considered to be unworkable. For example, outside of military applications, the use of formal methods is thought likely to involve excessive costs and be difficult to meet because of a lack of experienced staff.

4099 The work of the HSE in developing and promoting its guidelines for programmable electronic systems was regarded to be the right approach. But, the guidelines were thought to inhibit the implementation of safety related computer control systems because:

- of the rigours and therefore the costs that they imposed;
- they were ambiguous, and therefore:
  - require interpretation;
  - it becomes difficult for users to distinguish a bona fide supplier from a cowboy;
- there were no similar guidelines for mechanical or solid-state relay systems which means that they can more easily satisfy safety inspection and cost less to implement.

4100 The evidence of the study is that, whilst there was a generally understood need amongst suppliers to seek some generic, cross-application guidelines for SRCCS design and use, there was no agreement on the appropriate way forward and some doubts about the content of the recent HSE guidance.

(c) European developments

4101 In general terms, awareness of the likely impact of the Single European Market was poor. Only one SRCCS supplier gave any indication of being geared up to meet European market conditions and this was probably because it was the UK subsidiary of a major European company. It was generally felt that competition for SRCCS from suppliers based in Germany and France especially would increase. The rather sanguine response of UK based suppliers of SRCCS to this probably reflected that in the majority of cases their sales were not European.

4102 Moreover, the markets were characterised by close user-supplier relationships and repeat sales within niche markets. The increased demand for SRCCS tended to come predominantly from existing users (non-users appeared unlikely to be a source of demand in the short to medium term) and appeared to have been met by existing suppliers. Moreover, the technology embodied in the SRCCS being supplied to UK users was, for most applications, regarded by some of the suppliers as competitive and for some applications, eg offshore oil and gas, UK technology was considered superior. These points were reinforced during the focus group discussions where it was thought that the current favourable UK position derived from the fact that it was more safety conscious than the rest of Europe, ie the UK was currently ahead in its implementation of best practice. There was however no clear indication of the extent of implementation of best practice in the UK compared with other countries within Europe.

4103 The absence of a coherent picture emerging from suppliers to the UK market and from European trade associations on developments in the European market is not surprising given the fragmentation of the SRCCS market into self-enclosed niches on the one hand and into more general control systems on the other. However, with the limited information available it is still possible to hazard some guesses on potential developments in Europe:

- competitive advantage may well accrue to those suppliers who are able to transcend application specific boundaries whilst maintaining the integrity of their systems;

- this seems more likely to be achieved where there exists a recognised, third party testing and validation process especially for relatively low value products with significant sales potential (and thus where the costs of such validation will be marginal);

- this is likely to characterise the discrete and process manufacturing market (including the vehicle sector) rather than the energy, transport or defence manufacturing sectors where the unit cost of systems are higher and application specific standards exist or are emerging;

- the German suppliers are perceived by their competitors to have competitive advantage for the above reasons - traditional strength in discrete manufacturing combined with the TÜV system - and the French are working on creating international standards for the same competitive reasons;

- the UK suppliers do not have access to an equivalent UK system and appear to be more inclined to stay within their specialised niche markets and to extend their overseas presence outside Europe.

V Market impediments and areas for possible action

Identification of market impediments

501 Underpinning the terms of reference for the study was the aim to depict the market for SRCCS in order to establish whether the market was working efficiently in delivering safety to reduce risks to acceptable levels and to do so in the most cost-effective way. If it is not doing so, and there are identifiable market impediments, action may be appropriate for the industry (suppliers and users), government and the support infrastructure to enable the market to work more efficiently in future.

502 Market impediments will clearly be of particular importance where the costs and benefits perceived by the market agents (eg users, suppliers, intermediaries) in opting for safety systems differ from those facing society as a whole. In such circumstances, safety levels acceptable to society may be compromised by the operation of the market.

503 However, market impediments will also be of significance where socially acceptable levels of safety are achieved but only with an inefficient use of resources, ie with less than optimal safety solutions. For example, the market tends to operate by decentralising decisions to individual enterprises and this generally leads to efficient resource allocation. However, it is possible that this can give rise to insufficiently high degrees of standardisation to yield the combination of scale economies and competition which generates efficiency in systems' specification and operation.

504 The categories of potential market impediments likely to be of particular relevance to the market for SRCCS according to this study are, indeed, those identified in the SafeIT documentation. Thus the market may not work effectively where:

- there is poor awareness amongst potential users of:
  - standards relating to safety systems;
  - safety system options; and
  - appraisal, especially best practice, techniques

such that buyers have different levels of information about the options in the marketplace (differences both between buyers and between buyers and sellers) and that less than optimal decisions are likely to be made;

- high up-front costs and other entry barriers into markets reduce the incentives for suppliers to explore potential technical interrelatedness and economies of system scale (sometimes called network integration benefits) between SRCCS applications;
- the nature, scale and growth of technological change are such that, whilst it could transform production and consumption possibilities, nevertheless individual firms operating in niche markets may not be able to appreciate these possibilities.

505 Each of these market impediments is assessed in subsequent parts of this section with regard to the particular aspects of the SRCCS markets.

Information and awareness

506 Amongst the users of safety systems, whether or not computer controlled, the survey recorded a significant lack of information and awareness on SRCCS. The extent and nature of this is summarised in Table 11:

- only about 30% of all sites using a non-SRCCS safety system were aware of any SRCCS - awareness was much lower amongst small sites (18%);

- less than 10% of non-SRCCS user sites were aware of any SRCCS standards and only about one half of SRCCS users were aware of such standards;

- users in the energy and transport sectors were far more likely to be aware of SRCCS standards;

Table 11: Awareness amongst safety system users and their information needs: user survey (% sites)

(a) % of users of non-SRCCS who were aware of SRCCS:

	%
All sites	29
• small	18
• medium	31
• large	39

(b) % of users of SRCCS and non-SRCCS who were aware of SRCCS standards:

	%
All sites	21
• SRCCS users	53
• non-users	8
Energy	65
Transport	55
Discrete manufacture	14
Process manufacture	12
Small	10
Medium	16
Large	34

(c) % of users of SRCCS and non-SRCCS who considered:

	Information on SRCCS Unavailable or Difficult to Find	More Information was Needed	SRCCS Club to be of Interest
All sites	36	35	26
• SRCCS users	43	41	47
• non-users	34	33	18
Energy	39	37	57
Transport	28	38	38
Discrete manufacture	38	33	20
Process manufacture	36	36	24
Small	33	22	20
Medium	36	38	28
Large	41	42	30

(d) % of users of SRCCS and non-SRCCS wanting more information who wanted:

	General Information	Specialist Technical Information Related to Sector
All sites	53	17
• SRCCS users	45	19
• non-users	56	16
Energy	56	17
Transport	45	27
Discrete manufacture	56	15
Process manufacture	49	17

Note: Small sites employ 100 and under; medium between 101 and 500; large more than 500. Where no size breakdown is provided it is because no significant differences emerged from the survey.

- about one-third of all sites (and a higher proportion - 43% - of SRCCS users) found information on SRCCS was difficult or impossible to find and wanted more;

- general information on SRCCS rather than sectorally specialised or technical information was in demand from those wanting more information (mainly because of the absence of generic marketing) - this being the case amongst SRCCS users as well as non-users and amongst all sectors; and

- one quarter of all sites (and nearly half SRCCS users) expressed an interest in joining an SRCCS club or association, the strongest interest coming from the energy and transport sectors.

507 The lack of knowledge amongst users of non-computer controlled safety systems about SRCCS, particularly amongst the small sites, suggests that inadequate attention may have been given to the range of safety system options available. This observation combined with the study's findings on limited use of risk and hazard assessment and requirement specification procedures indicates that there is good reason to believe that less than optimal system solutions are being deployed.

508 The low level of awareness of SRCCS standards amongst non-users is unsurprising given their general lack of awareness of SRCCS. But the relatively low awareness of standards within sectors where penetration and awareness of SRCCS is otherwise high suggests that the standards either do not exist or have not been promulgated widely.

509 From the user perspective - even from the more mature user sectors of energy and transport - there was a clearly expressed need for more information of a generic kind and in a collaborative context (eg an association or club.). This is strongly suggestive of a situation in which users (and suppliers) may have become locked into sectoral or even application specific solutions which could be less than optimal on the wider view.

Market structure and integration

510 The analysis from the study indicates that safety is often not regarded as a primary feature of control systems, or at least not marketed directly as such. This, combined with the lack of information and awareness already highlighted, leads to a tendency for market competition to lock suppliers and users into historically and sectorally determined solutions and standards.

511 Thus, the study identified users of non-computer controlled safety systems of whom only one quarter to one third had even heard of SRCCS and just a handful of whom intended to install such systems. Amongst SRCCS users the study identified distinctions between:

- specialist system users where safety was a primary purpose (notably in transport) and other users where safety was integrated with, and to some extent subsidiary to, other control purposes (most significant in discrete manufacturing); and
- separate categories of users and suppliers in:
  - oil and petrochemicals;
  - power generation;
  - transport;
  - defence and aerospace engine control;
  - naval command and control;
  - instrumentation and testing; and
  - general industrial.

512 The impression of discrete, niche markets was reinforced by evidence of:

- limited marketing efforts beyond the existing customer base;
- a reliance on close and continuous supplier-customer relationships;
- the existence of sectorally specific standards; and
- an approach to system specification, installation, operation and maintenance which could be characterised as "do-it-yourself".

513 On this evidence about market conditions, there is the risk that the market may not facilitate development of, and access to, increasing returns available from benefits to:

- the supply-side from transferable technologies and best practice, more effective utilisation of scarce resources and improved system design facilitated by more efficient interfaces between components; and
- the demand-side where the gains to users increase more than in proportion to the increase in the number of users in the network.

514 There is the further risk under such conditions that suppliers and users could become locked into the "wrong" standards and that, through unregulated market forces, one variant (and not necessarily the best) amongst the available standards could emerge as the de facto standard for the sector or even more generally.

515 It is inevitably difficult to identify precisely what might or should have happened that has not. But there are particular aspects of network integration where there is some evidence that potential benefits may exist in stronger integration of the SRCCS market, notably in training and standards.

Training

516 One characteristic of "locked-in" markets is that they will rely on in-house training to a significant degree and thus there may be missed opportunities for third party, external provision of training. The benefits of the latter could be those of scale and cross-disciplinary and cross-sectoral transfer of experience.

517 The evidence of the study at the time it was undertaken did not suggest that there were any general recruitment problems. This may be attributable to the economic recession and otherwise more concern might have been expressed about the availability and content of external training.

518 As it was, our study found that:

- about 45% of users of non-computer controlled safety systems wanted more formal training in the safety area (somewhat more strongly expressed in manufacturing);
- a similar proportion of users of SRCCS (and of specialist SRCCS) wanted more formal training in the use of SRCCS - much the same proportion in all sectors;
- the reasons for the demand for more formal training amongst SRCCS users were:
  - to keep up with technological advances (35%);
  - to rectify lack of, or incomplete, understanding of the systems (30%); and
  - to ensure safe and efficient running of the plant (17%).
- for the specialist SRCCS users the main reason for more formal training was to keep up with technological advances (64%); and
- many suppliers saw advantages in increasing the integration of safety related issues into higher education courses and in broadening the latter to cover modelling, data structuring, design methods and teamwork.

519 Although this evidence suggests the existence of potential demand for training, it should be emphasised that the overall evidence from users, suppliers and training providers is that this demand has not so far translated into a significant demand for third party sourced training.

Standards

520 The pattern of awareness and use of standards related to SRCCS was much as would be expected in a fragmented market with certain segments more mature in SRCCS use. Thus, the standards tend to be specific to certain applications and sectors. Table 11 demonstrates that awareness amongst users was characterised by:

- generally low awareness of SRCCS standards amongst users of all types of safety systems;

- very low awareness of SRCCS standards amongst users of non SRCCS;
- much higher awareness in the energy and transport sectors (about 60%) compared with manufacturing (just over 10%).

521 The practice on the supply-side appeared similarly fragmented and uncertain. There was a variety of standards being applied and some moves towards more generic standards relating to quality certification (eg ISO 9000) but less emphatic commitment to general safety related standards and guidelines (eg the IEC standards under development and the HSE PES guidelines).

522 The view of suppliers was that there was a need for agreement on safety system standards and guidelines which:

- were international and specifically European;
- recognised the integration of the safety element in the general control function;
- could be associated with quality certification and/or implemented by a certifying authority for safety systems as a whole;
- emphasised guidelines rather than recommended standards, based on a "goal oriented" approach rather than a prescriptive methodology.

523 The market has reached a point in its development where there is a window of opportunity which the DTI and other public agents could use to exercise influence over market developments through training and standards.

Dynamics of technological change

524 There are some areas of technological development (and computer technology is one) where advances are so rapid and widespread that they transform production possibilities, sending shock waves through the market structure. It is our conclusion from the study that the market for safety systems, including SRCCS, is approaching this phase of development:

- the cumulative output (on which scale economies depend) of SRCCS is tending to grow more rapidly each year;
- just over one-third of all safety system user sites in the survey considered that the requirement for SRCCS would increase over the next two to five years - even one quarter of users of non SRCCS thought this;
- the main reasons for the expected increase were technological advances and diffusion and new legislation, standards or guidelines;
- the suppliers considered that the SRCCS imperative would be enhanced over the short to medium term by:
  - customer and liability issues;
  - public environmental awareness;
  - energy consumption concerns;
  - productivity (concern about down-time) and
  - loss of life, especially relative to competitor countries;
- the market was thought likely to open up further in the medium term with new entrants coming from:
  - increased international competition;
  - suppliers to the defence sector seeking civil applications.

525 If, indeed, the market is moving into this acceleration phase, it should be for consideration whether the infrastructural foundations (especially in training provision and standards/guidance promulgation) in the UK are adequate enough for its industries to compete for the substantial commercial opportunities which may arise later across existing niche markets and in new markets.

Possible actions

526 The programme of interviews with suppliers and infrastructure organisations carried out during the study drew out various observations on the actions which respondents thought desirable. Whilst there was no consensus on specific initiatives, there was general agreement on the following areas for action:

- awareness generation and information provision;
- standards and guideline specification and promulgation;

- training standards and provision; and
- transfer of best practice.

APPENDIX A

MARKET STUDY TERMS OF REFERENCE

Market Study Terms of Reference

1 To quantify the market for safety-related computer controlled systems throughout the European Community and the European Free Trade Area to establish:

(a) the number of systems in use;

(b) the value of the equipment in the controlling programmable electronic systems; and, separately, the direct cost of the software components;

(c) the value of the overall systems being controlled.

2 To analyse the market by:

(a) sector;

(b) type of application;

(c) criticality of application;

(d) features of the controlling system ie size, architecture, methods and tools, languages, standards;

(e) type of companies involved as developers, procurers, assessors and users of controlling systems; and

(f) the relative strengths and weaknesses of the UK and the most active European countries in terms of technical development of the controlling systems.

3 To describe how the market for controlling equipment operates with reference to:

(a) specialisation within and across industrial sectors;

(b) unique sector techniques and techniques that are common to more than one sector;

(c) barriers to trade presented by existing standards or lack of them, and presented by market forces;

(d) skill shortages eg combinations of safety and software engineering skills;

(e) the critical mass (in terms of size and strength of industrial base) needed to maintain and enhance the UK position within the market ie what mass is needed, and how near is the UK to achieving it?

(f) the benefit gained from the use of computer control for safety-related systems in relation to its cost (value for money);

(g) the impact of other markets eg security systems market, on this market.

4 To quantify for the purposes of the market covered by this study the post-experience education and training situation in respect of:

(a) the number of software engineers requiring education and training;

(b) the typical age, experience and qualifications of potential trainees;

(c) the time and finance likely to be available for training; and to assess

(d) the value of having a formal qualification arising from the training;

(e) the likely trends over the next ten years.

5 To explain how the gathered data has been validated as representative and to explain how the analysis of that data has been validated.

6 To estimate how the market for safety-related programmable electronic systems might develop over the next ten years and identify factors which would maintain and enhance the UK position, taking account of developments in standards and technology, and market changes (single market in 1992). Factors to account for include:

(a) changes in size and criticality of systems being developed;

(b) impact of EC Directives and IEC Standards (training or regulatory burden);

(c) impact of developments in software engineering;

(d) possible convergence of a dependability market;

(e) possible fragmentation of the market and the nature of this fragmentation.

APPENDIX B

BIBLIOGRAPHY OF SOURCE MATERIAL

Bibliography of Source Material

ACARD:	"Software - A Vital Key to UK Competitiveness", HMSO, 1986
Anderson T. Lee P.A.	"Fault Tolerance - Principles and Practice", 1981
Bell R, Smith S.	"Functional Safety of Programmable Electronic Systems - An Overview of draft IEC International standards" 1990
Bell R.	"Framework for the Design and Assessment of Safety-Related Control Systems - HSE Guidelines", 1989
Bell R. Clark B.J. Ward G.R.	"International and European Standards for Safety-Related Systems", 1990
Bennett P.A.	"An Overview of IEC Working Group 9", 1990
Bennett P.A.	"The March Towards Standards in Safety-Related Systems", 1990
Bloomfield R.E.	"SafeIT - 1: Overall Approach", 1990
Bloomfield R.E. Brazendale J.	"SafeIT -2 - Standards Framework", 1990
British Robot Association	"Robots and Flexible Automation for the Efficient Factory", British Robot Association Conference, 1984
British Robot Association:	"Robot Facts 1989", BRA, 1989
Brazendale J.	"A Framework for Achieving Safety-Integrity in Software", 1990
Centre for Software Engineering	"A Study of the Computer-Based Systems Safety Practices of UK, European and US Industry", HMSO, 1989
Clark B.J. Ward, G.R.	"The Application of Programmable Electronic Systems to the Control of Machinery - the Scene in Europe:, ACOS Workshop, 1990
Dale C.J. Foster S.	"The Development of Techniques and Safety and Reliability Assessment: Past, Present and Future", Safety and Reliability Society Symposium, 1987

DTI/SERC — Joint Framework for Information Technology: "Information Engineering Advanced Technology Programme"

DTI — "The Single Market - Machinery Safety", 1989

DTI — "The Single Market - New Approach to Technical Harmonisation and Standards", 1989

EEMUA — "Safety Related Instrument Systems for the Process Industries, Engineering Equipment and Materials Users Association", 1989

Finnie B.W. — "Introduction of New Methods for Assuring Safety into the Software Development Process" Critical Systems Series, IEE, 1990

Finnie B.W.
Johnston I.H.A. — "Practical Experience in the Assessment of Existing Safety Critical Computer Based Systems", SAFECOMP90, 1990

Finnie B.W. — "Design for Safety" Critical Systems Series, IEE, 1990

HSE — "PES - Programmable Electronic System in Safety-Related Applications (1 - An Introductory Guide, 2 - General Technical Guidelines)", HMSO, 1987

IEE, BCS — "Software in Safety-Related Systems", IEE, London 1989

IEE — "Requirements Capture and Specification for Critical Systems", Critical Systems Series, 1989

IEE — "Software testing for Critical Systems", Critical Systems Series, 1989

Institution of Nuclear Engineers — "International Conference on Control and Instrumentation in Nuclear Installations", INE, Glasgow, 1990 (various papers)

Jesty P.H.
Buckley T.F.
Hobley K.M.
West M. — "Drive Project V1051 - Procedure for Safety Submissions for Road Transport Informatics"

Johnston I.H.A. — Critical Computer Based Systems", SAFECOMP90, 1990

Martin B.R.
Wright R.I. — "The Thorp Approach to Safety Control", Safety and Reliability Society Symposium, 1987

McDermid J.A.	"Skills and Technologies for the Development and Evaluation of Safety Critical Systems". SAFECOMP90
Meffert K.	"Classification of risks in the event of malfunction of control systems: explanatory note on the application of DIN V 19 250", BIA Handbook, 11th edition, Erich Smidt Verlag, Bielefeld, 1989
Millward J.	"Systems Architectures for Safety Critical Automotive Applications", Critical Systems Series, IEE, 1990
Moon A.	"Vehicle Control Systems - Reliability Through Simplicity", Critical Systems Series, IEE, 1990
National Association of Lift Makers	"Programmable Electronic Systems in Safety Related Applications", (revised edition), NALM, 1990
Pearson J.	"Proposed HSE Guidelines on Emergency Shutdown Systems", 1990
Rata, Jean-Marie Albert	"The Work of the Technical Committee on Safety, Security and Reliability of Industrial Computer Systems", Safety and Reliability Society Symposium, 1987
Reed W.	"Safety Critical Software in Traffic Control Systems", Critical Systems Series, IEE, 1990
Thomas M.	"The Role of formal methods in developing safety-critical software". Critical Systems Series, IEE, 1990
Viswanadham, Sarma, Singh	"Reliability of Computer and Control Systems", 1987
Ward G.R. Clark B.J.	"Implications for Safety in Machinery Control Systems", 1990
Ward R.S.	"The Design of Safety Critical Remote Control Systems for Mining, 1990
Young R.E.	"Control in Hazardous Environments", 1982

APPENDIX C

ISSUES ARISING FROM THE DESK RESEARCH

PREPARED BY

COOPERS & LYBRAND
in association with
SRD (AEA TECHNOLOGY)

Issues Arising from the Desk Research

Contents

Issues Arising from the Desk Research

Identification of Issues

The objective of this appendix is to summarise, on the basis of a review of literature gathered, the issues that appear to be of most importance in the area of the use of computer technology to control machine functions, assure their safe operation and protect against hazard.

The issues identified as being important are grouped under the following headings:

(i) Definition of Safety Related Systems;

(ii) Controls:

- good practice;
- standards;
- certification;
- cost effectiveness;

(iii) Human Factors:

- human/machine interface;
- training and education;

(iv) System Delivery Life Cycle:

- hazard analysis;
- requirements definition;
- verification and validation/testing;

(v) Future Trends

(i) Definition of Safety Related Systems

To a certain extent it appears that there is some 'fuzziness' associated with what comprises a safety related system. A variety of terminology exists to express a safety related system and each substitutes the other eg, safety critical, 'safety related', 'safety system', 'safety classified' all of which are without a clear explanation. For example, safety systems may consist of the 'protection system' and the 'safety actuation system' while the safety related system can be used to refer to the 'monitoring system' providing operators (human or computer) with information to enable safety decisions to be made.

Organisations' understanding of safety systems in their particular operating environment needs to be established. This is likely to be important for two reasons:

- it establishes common ground between the interviewer and respondent and whether the interview should proceed to obtain more information as a greater level of detail; and

- indicates how respondents perceive the technical and functional limits of safety systems applications.

It is suggested that the IEC65A standard definition of related system be used. We propose to simplify the issue as much as possible by using the following working definitions:

- **Programmable Electronic Systems (PES)**

  A system based on a computer connected to sensors and/or actuators on a plant for the purpose of control, protection or monitoring.

- **Safety Related System**

  A system that controls and monitors any machines and processes and which is designed to prevent hazards and dangers to human life, and health and the environment.

- **Protection Scheme**

  A system designed to respond to conditions on the plant, which may be hazardous in themselves or if no action were taken could eventually give rise to a hazard, and to generate the correct outputs to minimise the hazardous consequences or prevent the hazard.

(ii) Controls

Good Practice

The issue of what constitutes 'good practice' is widely discussed. The literature suggests that good practice varies depending on the particular application/industry sector under consideration. Nevertheless, it is clear that 'good practice' will comprise a number of essential and logical steps which should be approached systematically to ensure the safety system or safety-related system will be effective. The core elements comprise:

- **hazard analysis** - to identify risks, risk limits and risk parameters
- **specification** - correct and in a formal specification language understandable by users, suppliers and external parties involved with the safety-related system;
- **verification** - 'are we building the thing right';
- **validation** - 'are we building the right thing'; and
- **operation and maintenance** - continuous checking to maintain efficient and effective functioning.

Overall, the design and implementation of a safety related system, together with software development, have been combined into a 'safety lifecycle' (IEC SC65A WG10) as shown in Figure 1.

This 'safety lifecycle' framework is still in draft form but is regarded as a good basis for engineering quality into safety related systems.

There is a suggestion, however, that such a framework is, in itself, insufficient and that the framework needs to be managed effectively. The adoption of BS5750 (ISO9000) has been suggested as a means of being able to attain a good general level of quality assurance; as a minimum in order to produce high quality documentation at all stages in the lifecycle, and the develop an adequate quality management control system.

Standards

An issue identified in reviewing the literature is that there is no single or generic standard for safety-related systems. Various application sectors have tended to develop their own specific standards and these have often been evolved with little reference to any other application - sector standards. As computer control devices have been incorporated into safety related systems there are now attempts to develop a generic approach, applicable across all application sectors. Similar moves are also occurring at an international level in terms of harmonising standards within and outwith the European Community. The objective here is to avoid cost penalties from diverging domestic and international standards.

The literature also highlights two further issues relating to the attitude of the software industry to standards. The software industry:

- is aware of the standards which exist but does not apply them in practice because they are regarded as unnecessary or too costly; and

- see existing standards as sufficient.

Certification

A suggestion to introduce a form of certification system for safety-related systems has been made. This is to ensure that a minimum level of competence is achieved by individuals and organisations operating in the field. While certification itself does not guarantee safety it may go some way to ensuring that a common and recognisable professional practice exists. However in the existing situation, where a number of application-specific standards exists, it is argued that a certification programme may be costly to implement and patrol.

Figure 1 : A Safety Systems Lifecycle Model

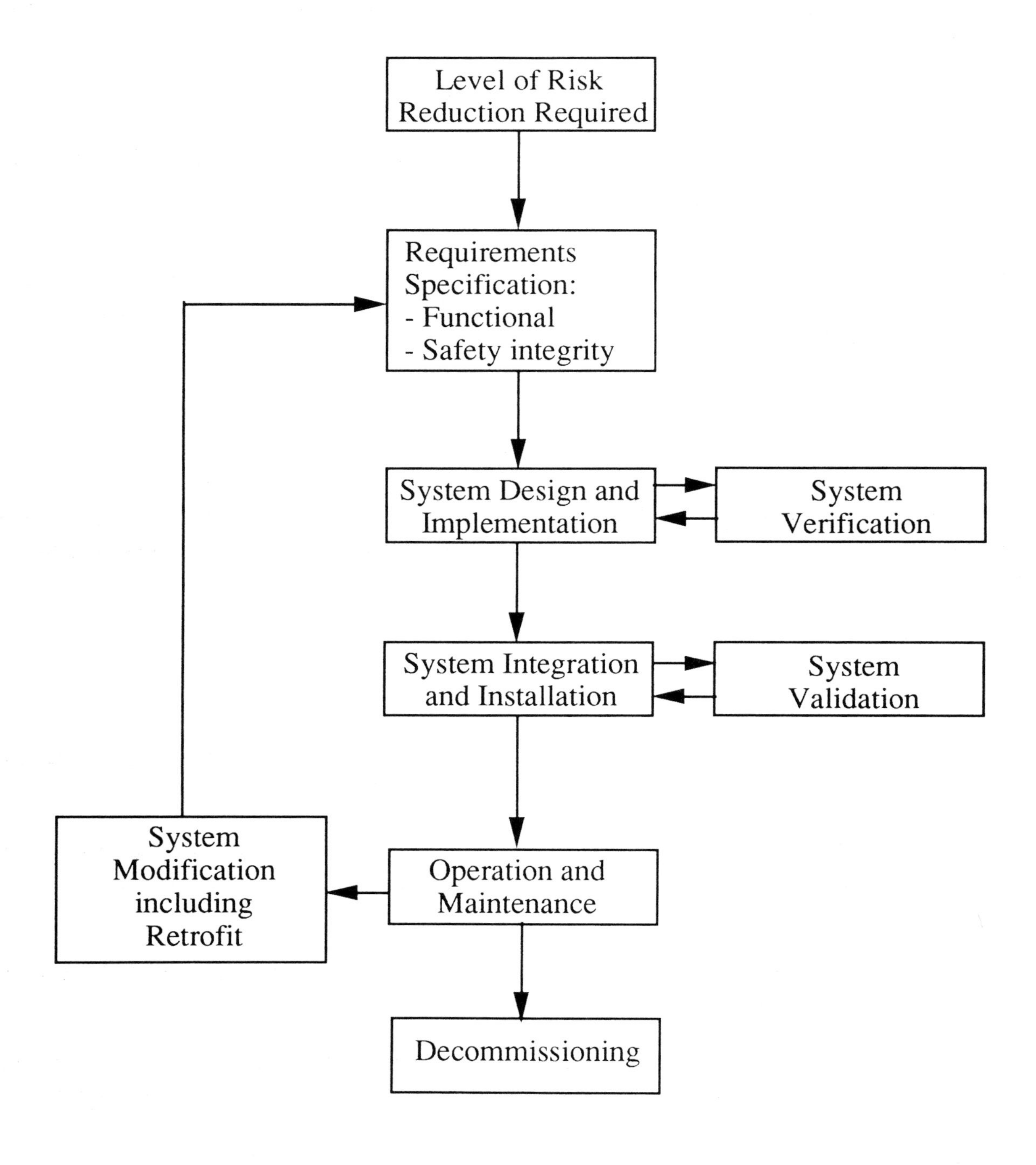

Cost Effectiveness

In terms of safety-related systems, the issue of safety is unlikely to be a safe/unsafe issue. Risk reduction has associated benefits and costs. Therefore the issue facing industry is most likely to be in terms of:

- what level of risk is acceptable or tolerable;
- at what point do the costs of further reductions in risk outweigh the benefits; and
- is it possible to quantify the cost/benefit equation.

It is suggested that developments in the area of safety-related systems, particularly in terms of standards, may impose a cost burden on industry that attention is focused too much on safety and that it is not adequately quantified.

(ii) Human Factors

Human Machine Interface

The role of humans in the safety-related system is also seen to be important, not least because human error can also affect the integrity of the system. There are two key areas where human involvement is important:

- the human-machine interface; and
- the human involvement throughout the life of a system.

The role of the human in the safety system, where a computer is present but does not exercise total control, is to:

- **perceive the state of the system** - data from the system must therefore be clearly relevant to the function and requirements of the operator and be presented within the range of human sensory functioning;
- **interpret the data provided by the system** - based on a mental model of the system that the human has developed; and
- **make decisions** about how the perceived state of the system matches the intended state and take appropriate action, where necessary, to attain the intended state.

Consequently, the physical design and construction of the human-machine interface is an important aspect of the safety and reliability of that system. It should therefore allow for the rapid and accurate assimilation of information and permit the operator to act upon the system through the controls, as intended.

Inclusion of software in the system does not alter the fundamentals of human-machine design. Rather, it does permit additional functions that, if appropriately used, can enhance the performance of both the operator and the system.

At each stage in the 'safety lifecycle', there will be human involvement which is critical to the overall integrity of the system. The human element is an important component in the specification and design of safety systems etc. As such, it is a potential source of error. This reinforces the requirement that all those associated with the design, production, implementation, maintenance and management of the system are adequately trained and work to prescribed quality assurance procedures.

In many instances, therefore, effective use of safety-related systems will require an element of cultural change; the development of a 'safety-oriented' attitude to complement existing engineering, team and management skills.

Training and Education

Given the importance of the human in the safety-system, a number of interesting features in terms of training and education emerge from the literature:

- there is a lack of convergence of opinion on staff qualification requirements;
- the discipline of safety engineering, especially in software, has not yet agreed to a basic set of skills required or detailed techniques of design and analysis; and
- there are no widely agreed standards for education in safety engineering and software.

More disturbing perhaps is that some of the proposed safety guidelines and standards appear to assume that those working in a safety related engineering environment (including software) will be trained to a standard set by the profession. The draft guidelines set out in IEC SC 65 A WG9 for example is vague on the subject of training and education, suggesting that the qualifications issue will require interpretation for particular industries.

In addition, where the human has a role in the operation of the safety system adequate training will be important to ensure the system works effectively.

There is, however, little to suggest that training is not important; more rather than less is the general message. However, the literature does not suggest that there is any shortage of skills in the safety related systems field.

For humans in the system, the areas where training to develop skills are seen as vital are:

- the acquisition (and reacquisition) of knowledge of the application domain; and
- improving interpretation, judgemental and decision making capabilities.

For humans involved in the specification, design side etc the areas where attention should be focused are:

- the acquisition (and reacquisition) of knowledge of the application domain;
- ensuring adequate experience through the use of a variety of environments, systems, techniques etc;

- training in the use of formal languages, techniques and tools of analysis; and

- training in the use of quality assurance procedures.

(iv) System Delivery Life Cycle

Hazard Analysis

There is a consensus in the various guidelines and draft standards that one of the important stages during the design and implementation of safety systems is that of hazard analysis ie an assessment of the potential hazards with which systems must cope in their operating environment. A number of hazard analysis techniques exist and have different strengths and weaknesses depending on the application in question. It is not clear how widespread this form of analysis is in practice. The question of integrity level requirement logically follows from the process of hazard analysis and assessment. It is an assessment of the extent to which the identified hazard is unacceptable or not. The acceptable integrity level of the safety related system then drives the required functions that are required to achieve a safe state and therefore specify the hardware and software architectures.

The draft guidelines that exist suggest a variety of integrity levels. DEFSTAN 00-56 for example leads to two levels of integrity; functions that are safety critical and those that are not. In contrast, IEC SC65A WG9 assumes there are five integrity levels while IEC SC 65 A WG10 suggests the number of integrity levels will be application dependent. The implication from these IEC draft standards are that it is possible to develop safety systems to one of a number (5) of different levels of assurance and that this is measurable, either quantitatively or qualitatively.

Requirements Definition

Problems are seen to arise because of communication difficulties between the individuals associated with the introduction of safety related systems. Those involved with the design and installation of the safety related system may not be intimately familiar with the specific requirements of the application. A critical issue is therefore that users (eg plant engineers) correctly and accurately identify and, more particularly, document the requirements that the safety related system must satisfy.

There are a number of areas which the literature highlights as being important:

- there does not appear to be a generic approach to specifying requirements;

- a variety of specification techniques have been developed but are only used in a limited number of sectors (eg defence, avionics and nuclear);

- specification techniques are frequently developed by academia rather than industry;

- there is no common agreement at academic or industry level on the use of specification techniques;

- the specification process is regarded as time consuming, costly and requires the training of suitably qualified individuals;

- there is no fully integrated set of tools for complete development from specification through to verification, operation, modification and maintenance;
- documentation is frequently compiled in natural language which can be inadequate or ambiguous for system designers or software engineers;
- documentation, particularly for software, is often constructed in a way which does not permit an easy audit and review and therefore presents problems if certification is required.

Verification and Validation/Testing

The basic issues here are:

- the verification and validation of specifications;
- does the system, executing within its operating environment, perform its safety function correctly on each occasion; and
- is it possible to assure the safety system, particularly the software component, against design faults.

The focus of attention in the literature is therefore directed at determining and assessing the various tools, techniques and developments used to provide a satisfactory answer to the questions: 'are we building the system right' and 'are we building the right system'.

Traditional methods of software specialists have been to develop methods of developing/purchasing error free software or proving it to be error free. The tools and techniques for this have improved eg through planned and statistically valid testing (although this will not be able to identify an accumulative failure); arithmetic testing, concurrent processes, identification and inspection of software discrepancies, and expert witness qualification.

What is clear, however, is that the verification and validation procedure, while important to the safe operation of the system, can be costly. There is therefore a requirement to adopt a cost-effective approach - it is more than just a safe/unsafe argument.

(v) Future Trends

The literature identifies a number of key trends in relation to safety-related systems. These can be broadly grouped under two headings: technology and standards.

Technology

The main trends identified here are:

- the increased use of 'application orientated technology';
- development of the 'theory of composition' - integration of 'safe' components;

- development of ‘complete verified systems’ - the integration of the systems development cycle within one formal analytical framework;
- the application of the ‘theory of composition’ to maintenance - to avoid re-verification;
- the increased use of Application Specific Integrated Circuits (ASICs) to reduce system complexity.

Standards

The main trends identified here are:

- the production of generic guidelines for safety related systems;
- progressing the development of guidelines and standards internationally - the main objective being international standardisation.

APPENDIX D

SUPPLY SIDE DISCUSSION GUIDES

Discussion Guide for Suppliers, Regulatory Authorities, Academic Research Departments and Infrastructure Bodies

A Your Business

1. Size
   - turnover
   - employees
   - offices - UK, other

2. Description
   - sales by sector
   - sales by application/product
   - sales by country

3. Safety Related Computer Control (SRCC) work
   - definition of SRCC
   - history of SRCC work
   - % of total business
   - sales by sector
   - sales by application/product
   - sales by country
   - future development

4. Membership of consortia, associations etc

B SRCC Market

1. Market characteristics
   - size of market in UK/abroad
   - application product areas in UK/abroad
   - main competitors/key players in UK/abroad
   - supply chain
     - to user direct
     - to intermediary (user indirect)
     - division of own company (internal user)
     - other

2. Corporate SRCC strategy
   - dedicated sales and marketing
   - marketing approach
   - sales life-cycle
   - competitive position
     - mkt share
     - comparison with competition

3. Key issues
   - problems
   - constraints
   - opportunities
   - UK market position

C SRCC System Delivery

1. Product description
   - type(s)
   - application area(s)
   - package/tailored
   - cost
   - position in market

2. User requirements
   - who defines
   - how defined
   - problems

3. Development
   - use of software/hardware
   - use of tools/techniques
   - controls
   - testing
   - documentation
   - time/costs

4. Installation/Implementation/Maintenance
   - who
   - how long
   - provision of support
     - technical maintenance
     - user support

D Training and Education

1. Internal SRCC skills base
   - number of staff
   - qualifications required

2. Training provision
   - internal/external courses
   - course requirements
   - course content
   - costs/benefits

3. Recruitment
   - sources
   - problems/constraints
   - impact on business
   - potential improvements
   - recruitment and retention

4. User training
   - requirements
   - nature and extent
   - policy

5. Future skills requirements
   - type
   - number
   - qualifications
   - expected source
   - problems/constraints

E Standards and Legislation

- awareness
- use
- usefulness
- impact on business

F Information/Expertise

- sources
- comprehensiveness
- availability
- subject

G The Future

- impact of 1992
- standards and legislation
- technology
- use of SRCC systems
- human/environmental issues
- market share
- structure of supply/demand
- interest in joining SRCC 'club'[1]

[1] The discussion guide was developed prior to the establishment of the 'Safety Critical Systems Club'.

Discussion Guides for Training and Education Providers and Academic Departments

A. Your Organisation

1. Size
   - turnover
   - employees
   - offices - UK, other
   - applications

2. Description
   - of organisation
   - courses provided
   - training approach used
   - type of training (set course, bespoke)

B. Training Provision

1. Training provision
   - content/level/length
   - relevance to SRCCS (specific individuals/applications)
   - element of wider training package

2. Demand for Training
   - key customers
   - satisfaction levels
   - volume
   - value
   - growth
   - problems/constraints

3. Internal Skills Base
   - number of staff in SRCCS field
   - qualifications
   - requirements
   - source
   - own training
   - problems/constraints

4. Future
   - technology
   - standards
   - impact of technology and standards on skills
   - user requirements

C. Market for Training

1. Training provision
   - main suppliers
   - marketing approach
   - competitive position

APPENDIX E

TELEPHONE QUESTIONNAIRES:

- **DEMAND SIDE USER SURVEY**
- **NON-USER SURVEY**

Interviewer name..............................Date.........

SAFETY RELATED COMPUTER CONTROLLED SYSTEMS

DEMAND SIDE SURVEY QUESTIONNAIRE

13 May 1991

Good morning/afternoon, my name is .... from Benchmark Research. We are currently conducting a short study on behalf of the DTI into the use of Safety Related Computer Controlled Systems. These are systems that control and/or monitor machines and processes, with the specific remit to directly prevent both immediate and long-term danger to both the human population and the environment, including physical property. Would you mind answering some questions? Any information that you provide during our discussion, which should last about 15 minutes, will be treated in the strictest confidence.

Q.1 Firstly, could you tell me whether this site has implemented a safety related computer control system. That is to say, a system that controls and monitors any machines and processes and which is designed to prevent hazards and dangers to human life.

(This may be a dedicated computer system, designed to operate in a safety role only, or a general control system which has an in-built safety/protection system **and** is used in a specific safety capacity)

Yes	1	
No	2	GO TO Q.16
Don't Know	3	ASK TO BE TRANSFERRED TO SOMEONE WHO WILL BE ABLE TO HELP YOU. REPEAT THE INTRODUCTION.

Q.2a How many safety related computer control systems do you have?

....

FOR Q.2b TO Q.2h WRITE IN ANSWERS BELOW

Q.2b Please can you give me the names of the system(s)

Q.2c How many of this type do you have?

Q.2d Does this system protect against loss of life or disablement?

Q.2e Is the system dedicated or is it a general control system with an in built safety function?

Q.2f What was the approximate total investment for the system, including all internal and external costs?

Q.2g When was it installed?

Q.2h Is the system package or bespoke? (Both may apply)

Q.2b Name	Q.2c Number	Q.2d Loss of life Y/N	Q.2e Dedicated Y/N	Q.2f Cost	Q.2g Year	Q.2h P/B
1............	.....	.....	.....	.....	.....	.....
2............	.....	.....	.....	.....	.....	.....
3............	.....	.....	.....	.....	.....	.....
4............	.....	.....	.....	.....	.....	.....
5............	.....	.....	.....	.....	.....	.....

Q.2i From now on we will concentrate on the systems designed to protect against loss of life or disablement (Q.2d). For these systems specifically, what is the frequency of risk against which the computer system(s) protects?

(WRITE IN BELOW)

Frequent - Many time a year	1
Probable - Once per year	2
Occasional - During product lifetime	3
Remote - Unlikely	4
Improbable/Incredible	5

Name of system (Q.2b)	Frequency of risk
1......................	____
2......................	____
3......................	____

FOR THE FOLLOWING QUESTIONS CONCENTRATE ON THE SYSTEM WITH THE GREATEST FREQUENCY OF RISK (Q.2i), WHERE AN ACCIDENT WILL INVOLVE LOSS OF LIFE OR DISABLEMENT (Q.2d)

WRITE IN:..

GO TO Q.3a

IF 2 SYSTEMS HAVE THE SAME FREQUENCY OF RISK AND PENALTY FOR FAILURE , ASK:

Q.2J Which safety related computer control system do you regard as most important?

....

Why?

....

Q.3a Which of the following hazards is the system designed to prevent:

Operator Fatal Injury	1
Non-operator Employee Fatal Injury	2
Spillage of dangerous substances	3
Malfunctioning of a process system	4
Environmental Damage	5
Damage to Plant	6
Injury to the general workforce	7
Injury to the general public	8
Other (write in)	9

Q.3b What are the protective features of the computer related control system that prevent the hazard arising? (More than one may apply)

(i)	1
(ii)	2
(iii)..............	3
Other (Write in)	4

....

Q.4 (As appropriate) Could you tell me for what specific application or applications the system(s) is being used.

....

....

....

Q.5 What is the estimated total cost of implementing the system? Please include all internal and external costs.

£ ....

Q.6 From where did you obtain the system?

In-house created	1	**Who?**
Software House	2	....
Consultant	3	....
Hardware Company	4	....
Other (Write in)................5		

Q.7a Thinking more generally, which of the following factors do you consider to be important in the decision to select a safety related computer system.

	Important	Unimportant
This site adheres to accredited standards	1	2
The system is used in similar Industries	1	2
The enhancement of the Company's reputation in the marketplace through using the system	1	2
The need to minimise potential risk	1	2
The process demands the safety system	1	2
Cost/benefit analysis	1	2
Compliance with HSE requirements	1	2
Recommendation from third party	1	2
Cost of training in system use	1	2
Degree of hazard controlled	1	2
In-built in the process machinery	1	2
Other(specify)...............	1	

Q.7b Which of the following features or factors do you consider to be important when acquiring a specific safety related computer controlled system?

	Important	Unimportant
Level of maintenance	1	2
Ease of use	1	2
Level of external support required	1	2
Level of integrity	1	2
Fault tolerance levels	1	2
Functionality	1	2
Ease of installation	1	2
Cost effectiveness	1	2
Product Guarantee	1	2
Reputation of the Supplier	1	2
Ability of system to meet exact needs	1	2
The price of the System	1	2
The cost of maintenance of the system	1	2
Other(specify).............	1	

Q.8a Who specifies the type of system to be implemented?

		Who?
In-house experts	1	
External Consultants	2	....
External Software developers	3	....
Other (write in) ____________	4	....

Q.8b Were specification guidelines used to design the system?

Yes 1

No 2 (SKIP TO Q.9)

Q.8c Can you briefly describe them?
(Probe fully)

....

....

....

Q.9a Were specific techniques used to specify the system requirements?

Yes 1

No 2 (SKIP TO Q.10)

Q.9b Can you briefly describe them?
(Probe fully)

....

....

....

Q.10 I am now going to read out a list of problems or constraints that may have been encountered during the specification of the system. For each could you tell me if this was indeed encountered.

Lack of hazard assessment technique	1
Lack of appreciation of hazards	2
Lack of standardised procurement procedure	3
Difficulty in designing the system	4
Cost of designing the system	5
Timescales involved in designing the system	6
Inability to prescribe what the system should do	7
Inability to meet required standards	8
Lack of conformance testing	9
Lack of suitable hardware platform	0
Lack of in-house skills to design the system	X

What others? (Please specify)

....

....

Q.11 Thinking about the actual installation of the system could you tell me:

a) Who installed the system? ....

b) How long in months it took to become fully operational? ....mths

c) How long you expected to wait before it became fully operational?......................................mths

Q.12a Who undertook the testing and validation of the system?

....

Q.12b Were any specific tools or guidelines used to validate the system?

Yes	1	Which? ....
No	2	
Don't know	3	

Q.12c Approximately how long was the validation process?

.... (enter months)

Q.13a Moving on to think about the maintenance of the system, is this carried out by:

In-house experts	1	**Who?**
The original system supplier	2	....
A different third party	3	....
Other (write in) _______	4	....

Q.13b Would you say that the level of maintenance provided is:

Very satisfactory	1
Fairly satisfactory	2
Fairly unsatisfactory	3
Very unsatisfactory	4

Q.13c IF FAIRLY OR VERY UNSATISFACTORY ASK:

Why do you say that?

....

....

....

Q14a Is training provided at this site for hazard assessment?

Yes 1

No 2

Q.14b In which of the following areas of safety related computer controlled systems, has training been provided for employees at this site?

System specification	1
System design	2
System validation	3
System installation	4
System Use	5
System Maintenance	6
None	7 (SKIP TO Q.14g)

Q.14c Was the training:

Internal?	1	**Who?**
External?	2	....
Both?	3	....

Q.14d Approximately how many man days were given up for training?

....

Q.14e Approximately what level of investment was made in training for the system?

£................

Q.14f How useful would you say the training has been in regard to using the system?

Very useful	1
Fairly useful	2
Of little use	3
Not at all useful	4

Q.14g Who, in terms of job title, actually attended the training?
(Probe for all job titles)

....

....

SKIP TO Q.15a

Q.14h FOR SITES THAT HAVE NOT USED TRAINING

Why have you not undertaken any training on site related to the safety related control systems?

....

....

....

Q.15a Do you think that there should be more formal training in the use of safety related computer control systems?

Yes	1
No	2

Q.15b Why do you say that?

....

....

....

Q.16 Do you plan to implement any safety related computer controlled systems in the coming twelve months?

Yes 1

No 2 (SKIP TO Q.22a)

Don't know 3 (SKIP TO Q.22a)

Q.17 a) Which types of safety related computer control system will you implement? (More than one may apply)

b) How many of each will be installed?

c) Will the system be package or bespoke

d) Will the safety related computer control system guard against loss of life or disablement?

	A) Planned	B) Number	C) P/B	D) Y/N
....	1	...	...	...
....	2	...	...	...
....	3	...	...	...
....	4	...	...	...
Other (write in)	5	...	...	...

Q.17e Thinking now about the safety related computer system(s) that protects against loss of life or disablement For these systems specifically, what is the frequency of risk against which the computer system(s) protects?

(WRITE IN BELOW)

Frequent - Many time a year	1
Probable - Once per year	2
Occasional - During product lifetime	3
Remote - Unlikely	4
Improbable/Incredible	5

Name of system	Frequency of risk
1......................	____
2......................	____
3......................	____

FOR THE FOLLOWING QUESTIONS CONCENTRATE ON THE SYSTEM WHICH WILL HAVE THE GREATEST FREQUENCY OF RISK, WHERE AN ACCIDENT WILL INVOLVE LOSS OF LIFE OR DISABLEMENT

WRITE IN:...

GO TO Q.18

IF 2 SYSTEMS HAVE THE SAME FREQUENCY OF RISK AND PENALTY FOR FAILURE , ASK:

Q.17f Which safety related computer control system planned do you regard as most important?

....

Why?

....

CONCENTRATE ON THIS SYSTEM

Q.18 Could you tell me for what application or applications the system(s) will be used?

....

....

....

Q.19 Which of the following hazards is the system designed to prevent:

Operator Minor Injury	1
Operator Fatal Injury	2
Spillage/leakage of dangerous substances	3
Malfunctioning of a process system	4
Environmental Damage	5
Damage to Plant	6
Injury to the general workforce	7
Injury to the general public	8
Other (write in) ___________	9

Q.20 What is the estimated total cost of the system. Please include all internal and external costs.

£.............

Q.21 From where do you plan to obtain the system?

In-house	1	**Who?**
Software House	2	....
Consultant	3	....
Hardware Company	4	....
Other (write in) ________	5	

Q.22a Are you aware of any standards related to the use of safety related computer control systems for your industry?

Yes 1

No 2 (SKIP TO Q.24)

Q.22b Can you briefly describe what they cover?

....

....

....

Q.22c Do you apply the standards in practice?

Yes 1 (SKIP TO Q.23)

No 2

Q.22d Is this because standards are seen as:

too costly to implement	1
unnecessary for the application	2
unnecessary for the level of integrity	3
not corporate policy	4
other (write in) _______________	5

Q.23 Has the introduction of standards had a direct influence on the use of safety related computer controlled systems at this site?

Yes	1
No	2
Don't Know	3

Q.24 Do you think that the impact of new legislation in 1992 will:

require the company to move towards more standards	1
encourage the company to look at standards more seriously	2
have no effect at all	3
DO NOT READ OUT	
don't know	4

Q.25a Have you made definite plans for the advent of 1992, in terms of safety related computer systems?

Yes	1	
No	2	(SKIP TO Q.26)

Q.25b Can you briefly describe them?

....

....

....

Q.26a Over the next 2-5 years, do you expect your requirements for safety related computer controlled systems to:

Increase	1
Stay the same	2
Decrease	3

Q.26b Why do you say that?

....

....

....

Q.26c Over the next 2-5 years, do you expect your expenditure on safety related computer controlled systems to:

Increase	1
Stay the same	2
Decrease	3

Q.26d Why do you say that?

....

....

....

Q.27 What do you think will key issues over the next 2 to 5 years, regarding the use of the safety related computer control systems? **(Probe Fully)**

....

....

....

Q.28 What about the next 5 to 10 years? Do you anticipate your requirements for safety related computer controlled systems, compared to today, to:

Increase	1
Stay the same	2
Decrease	3

Q.29 Thinking more generally, which of the following sources of information do you use in order to obtain information on safety related computer systems and which do you find most useful?

		Most Useful
Research Papers	1	1
Case Studies	2	2
Trade Journals	3	3
Trade Bodies	4	4
Software Suppliers	5	5
Equipment Suppliers	6	6
Government Departments	7	7
Government Publications	8	8
Consultants	9	9
Word of Mouth	0	0
Training Establishments	X	X
Other (Write in)________	Y	Y

Q.30 Would you say that information on safety related computer control systems is:

readily available?	1
Difficult to find	2
Not available at all	3

Q.31a Do you feel that you require more information on safety related computer controlled systems, in order to enable the site to take advantage of their benefits?

Yes 1

No 2 (SKIP TO Q.32)

Q.31b What type of information do you need?

....

Q.32 Would you be interested in joining the Safety Related Computer Control Systems User Club? [1]

Yes 1

No 2

[1] This was asked prior to the establishment of the 'Safety Critical Systems Club'.

CLASSIFICATION DETAILS

Name of Respondent	....			
Job Title	....			
Company Name	....			
Address	....			
	....			
	....			
	....			
Telephone Number	....			
Turnover	....			
Number of Employees	....			
Offices Abroad	Yes	1	No	2
	IF YES: Where is HQ sited?			
	....			
Industry Sector	Discrete Manufacturing	1		
	Process Manufacturing	2		
	Energy	3		
	Transport	4		
SIC Code	....			
Main Business Area	....			

Interviewer Name.......................Date............

SAFETY RELATED COMPUTER CONTROL SYSTEMS

NON-USER STUDY QUESTIONNAIRE

July 1991

INTERVIEWER NOTE: USE THIS QUESTIONNAIRE ONLY FOR SITES IDENTIFIED AS BEING NON-USERS OF SAFETY RELATED COMPUTER CONTROL SYSTEMS, EITHER DEDICATED OR NON-DEDICATED.

Q.2a Do you plan to implement any safety related computer control systems in the coming 12 months?

YES	1	GO TO Q.5
NO	2	
Don't Know	3	

Q.2b Are you aware of any safety related computer control systems?

YES	1	
NO	2	GO TO Q.3a

Q.2c Have you ever considered using a safety related computer control system?

YES	1	GO TO Q.2e
NO	2	

Q.2d Why was a safety related computer control system not considered?

DO NOT READ OUT. PROBE FULLY.

The process is not perceived to be dangerous enough	1
There are no standards requiring a safety related computer control system	2
No cost benefit through implementation	3
Other(s)	4

....

....

....

GO TO Q.3a

Q.2e Why was a safety related computer control system not installed?

DO NOT READ OUT. PROBE FULLY.

The process is not perceived to be dangerous enough	1
There are no standards requiring a safety related computer control system	2
No cost benefit through implementation	3
Other(s)	4

....

....

....

Q.3a What do you think are the advantages of using a non-computer related computer control system?

....

....

....

Q.3b What type of safety related system do you use?

Manual	1
Hard-wire/electronic	2
Mechanical	3
Other	4

....

Q.4 What do you believe would encourage increased usage of Safety Related Computer Control Systems?

....

....

IF NO SYSTEM IS PLANNED GO TO Q.11a

Q.5 a) Which types of safety related computer control system will you implement? (more than one may apply)

b) How many of each will be installed?

c) Will the system be package or bespoke?

d) Will the safety related computer control system guard against loss of life or disablement?

	a) Planned	b) Number	c) P/B	d) Y/N
....	1	....	....	....
....	2	....	....	....
....	3	....	....	....
....	4	....	....	....

Q.5f Thinking **only** about the safety related computer system(s) that protects against loss of life or disablement, what is the frequency of risk against which the computer system protects?

Frequent - Many times a year	1
Probable - Once a year	2
Occaisional - During product lifetime	3
Remote - Unlikely	4
Improbable/Incredible	5

WRITE IN BELOW

Name of system	Frequency of risk
1......................	....
2......................	....
3......................	....

FOR THE FOLLOWING QUESTIONS, CONCENTRATE ONLY ON THE SYSTEM WHICH WILL HAVE THE GREATEST FREQUENCY OF RISK (WHERE AN ACCIDENT WILL INVOLVE LOSS OF LIFE OR DISABLEMENT)

WRITE IN...

GO TO Q.7

IF 2 SYSTEMS HAVE THE SAME FREQUENCY OF RISK AND PENALITY FOR FAILURE, ASK:

Q.6 Which planned safety related computer control system do you regard as most important?

....

Why?

....

CONCENTRATE ON THIS SYSTEM FOR THE FOLLOWING QUESTIONS

Q.7 Could you tell me for what application or applications the system(s) will be used?

....

....

....

Q.8 Which of the following hazards is the system designed to prevent:

Operator Minor Injury	1
Operator Fatal Injury	2
Spillage of dangerous substances	3
Malfunctioning of a process system	4
Environmental Damage	5
Damage to plant	6
Injury to the general workforce	7
Injury to the general public	8
Other (WRITE IN)______________	9

Q.9 What is the estimated total cost of the system. Please include all internal and external costs.

£........................

Q.10 From where do you plan to obtain the system?

In-house	1	**Who?**
Software House	2	....
Consultant	3	....
Hardware Company	4	....
Other (write in) ________	5	

Q.10a Why are you planning to install the system at this specific time?

....

....

Q.11a Do you provide any training related to the safety of the process?

YES 1

NO 2 GO TO Q.12

Q.11b Can you breifly describe this training?

....

Q.12 Do you think there should be more formal training in the safety arena?

YES 1

NO 2

Q.13a Are you aware of any standards related to the use of safety related computer control systems for your industry?

YES 1

NO 2 (GO TO Q.15)

Q.13b Can you breifly describe what they cover?

....

....

....

Q.13c Do you intend to apply the standards in practice?

YES 1 (GO TO Q.14)

NO 2

Q.13d Is this because standards are seen as:

Too costly to implement	1
Unnecessary for the application	2
Unnecessary for the level of integrity	3
Not corporate policy	4
Other (write in) _______________	5

Q.14 Has the introduction of standards had a direct influence on the use of safety related computer control systems at this site?

YES	1
NO	2
DON'T KNOW	3

Q.15 Do you think that the legislation in 1992 will:

Require the company to move towards more standards	1
Encourage the company to look at standards more seriously	2
Have no effect at all	3

DO NOT READ OUT

Don't know	4

Q.16a Have you made definite plans for the advent of 1992, in terms of safety related computer control systems?

YES	1	
NO	2	(GO TO Q.17)

Q.16b Can you briefly describe them?

....

....

....

Q.17a Over the next 2-5 years, do you expect your requirements for safety related computer controlled systems to:

Increase	1
Stay the same	2
Decrease	3

Q.17b Why do you say that?

....

....

Q.17c Over the next 2-5 years do you expect your expenditure on safety related computer controlled systems to:

Increase	1
Stay the same	2
Decrease	3

Q.17d Why?

....

....

Q.18 What do you think will be the key issues over the next 2 to 5 years, regarding the use of safety related computer control systems?

....

....

....

Q.19 Over the next 5 to 10 years do you anticipate your requirements for safety related computer control systems to:

Increase	1
Stay the same	2
Decrease	3

Q.20 Which of the following sources of information do you use in order to obtain information on safety related computer control systems?

Which **one** is the most useful?

	In Use	Most Useful
Research Papers	1	1
Case Studies	2	2
Trade Journals	3	3
Trade Bodies	4	4
Software Suppliers	5	5
Equipment Suppliers	6	6
Government Departments	7	7
Government Publications	8	8
Consultants	9	9
Word of Mouth	10	10
Training Establishments	11	11
Other (Write in) ________________	12	12

Q.21a Would you say that infomation on safety related computer control systems is:

Readily available	1
Difficult to find	2
Not available at all	3

Q.21b Do you feel that you require more information on safety related computer control systems, in order to enable the site to take advantage of their benefits?

YES	1	
NO	2	(GO TO 22)

Q.21c What type of information do you need?

....

Q.22 Would you be interested in joining the Safety Related Computer Control Systems Association or Club? [1]

YES	1
NO	2

[1] This was asked prior to the establishment of the 'Safety Critical Systems Club'.

CLASSIFICATION DETAILS

Name of respondent ....

Job title ....

Company Name ....

Address ....

....

....

....

....

Telephone Number ....

Turnover ....

Number of Employees ....

Offices Abroad YES 1 NO 2

IF YES: Where is HQ sited?

....

Industry Sector	Discrete Manufacture	1
	Process Manufacture	2
	Energy	3
	Transport	4

1980 SICC ....

Main Business Area ....

APPENDIX F

THE DEMAND SIDE ANALYSIS

PREPARED BY

COOPERS & LYBRAND DELOITTE
in association with
BENCHMARK RESEARCH LTD

The Demand Side Analysis

Contents

The Demand Side Analysis

Preface

1 There were two components to the demand-side analysis. The first was the sample survey. This was designed in part to facilitate the second component which was the estimation of the market for SRCCS and its key characteristics. However, not all of the survey's findings could be used in this way because of the limited sample size in some of the key areas of interest.

2 Consequently, the demand-side analysis is reported in this appendix in two sections. The first is a summary of the sample survey results with no attempt at generalisation to the population of SRCCS users or safety system users.

3 The second section of this appendix describes how we used our understanding of the survey sample, and the population from which it was derived, to come to conclusions about the characteristics of the market which the terms of reference for the study required us to address.

I Report on the demand side survey

Introduction

101 This part of the appendix reports the results of the survey of safety related systems users (the demand side). The survey was undertaken by means of telephone interviews around a questionnaire agreed between the DTI, their advisers, Coopers & Lybrand Deloitte and Benchmark Research. The fieldwork was carried out by Benchmark but this report of the findings was prepared by Coopers Deloitte.

102 The response rate (calculated as the number of positive responses as a proportion of positive responses plus refusals) was very high at 97%. It reflects the importance of, and the interest in, the subject matter. It also indicates the statistical reliability of the survey findings. However, it must be emphasised that what is reported in this appendix are the sample survey results. The translation of these into estimates of market size and growth requires further analysis which is described in Part II of the appendix.

103 Three sections follow this introduction. The next section examines the survey results from all sites, assessing awareness of safety standards and new legislation, expectations about increased requirements for safety systems, availability of information on such systems and the extent of use and intended use.

104 The third section considers the results from non-users of safety related computer controlled systems (SRCCS), identifying the types of control system they use, whether they have considered using computer controlled safety systems and with what results, and what their information and training requirements are.

105 The fourth section looks at SRCCS users and assesses their particular characteristics and requirements, drawing out any distinctions between specialist and other users. Specialist users are defined as those who use computer control equipment specifically to control for safety (rather than any other control purpose) and who rely on those suppliers identified from our supply-side interview programme to specialise in such dedicated equipment. This is recognised to be a more well-defined and narrower market than more general control equipment where safety is only one of the attributes.

All sites using safety control

106 Nearly 500 sites provided responses to the survey. Of these only 1 per cent (five sites) claimed not to use any type of safety control but over 30% did not know what type of safety control was in place on the site. Table 1 presents the distribution amongst the sampled sites of the types of safety control in use.

Table 1: Types of safety control in use (numbers of users)

	Total	Discrete	Process	Energy	Transport	Small	Medium	Large
		Manufacturing				Site size		
SRCCS users	141	39	41	41	19	16	60	55
• specialists	50	4	15	16	15	3	18	22
• other	91	35	26	25	4	13	42	33
Safety Control users	356	173	165	8	10	106	168	78
• hard-wire*	172	79	88	2	3	47	82	43
• mechanical*	164	74	85	1	4	38	85	41
• other*	158	76	76	1	5	43	78	37
• none	5	1	3	-	1	-	1	3
• don't know	109	58	43	5	3	37	48	21
Total	**497**	**212**	**207**	**49**	**29**	**127**	**228**	**133**

* More than one type of system in use at any one site.

NB: Small sites are those employing 100 and under; medium sized sites employ between 101 and 500; and large sites are those employing more than 500.

107 The penetration of safety related computer controlled systems (SRCCS) was highest in the energy and transport sectors and amongst the largest sites (84%, 66% and 41% respectively). This was particularly the case for the specialist SRCCS users. There were no marked distinctions between the use of hard-wire, mechanical and other safety controlled systems in each sector or site size band. Table 2 presents the penetration figures from the sample.

Table 2: Penetration (% of sample in each sector and site size band)

	Total	Discrete	Process	Energy	Transport	Small	Medium	Large
		Manufacture				Site Size		
SRCCS	28	18	20	84	66	13	26	41
Specialist	10	2	7	33	52	10	8	17
Hardwire*	35	37	42	4	10	39	36	32
Mechanical*	33	35	41	2	14	31	37	31
Other*	32	36	37	2	17	35	34	28
None	1	-	1	-	3	-	-	2
Don't Know	22	27	21	10	10	30	21	16

* More than one type of system in use on any one site.

108 Awareness of standards relating to the use of SRCCS was highest in the energy and transport sectors, amongst larger sites and amongst users of SRCCS. But there was a significant lack of awareness or knowledge amongst all sectors and site sizes - see Table 3.

Table 3: Awareness of SRCCS standards (% of all sites)

	Total	Discrete	Process	Energy	Transport	Small	Medium	Large	SRCCS Users	Non-users
		Manufacture				Site Size				
Aware	21	14	12	65	55	11	16	34	53	8
Unaware	48	50	52	35	28	47	52	43	46	49
Don't know	31	36	36	05	17	42	32	23	1	43

109 Of those who claimed to be aware of SRCCS related standards (some 100 sites), nearly one third cited the HSE 1974 regulations and about one fifth quoted non-specific British standards. Amongst these 100 sites, some 60% claimed that the introduction of standards had exercised a direct influence on the use of SRCCS - this included over 40% of those who were not currently using SRCCS! It is possible that many of the latter were reconsidering their position in the light of their awareness of the potential impact of new legislation in 1992. Some two-thirds of all those aware of standards were anticipating that 1992 legislation would cause them to take the standards more seriously or move to their adoption, but only about 20% had definite plans for the advent of 1992 changes.

110 Over one-third of the total sample of nearly 500 sites considered that the requirement for SRCCS would increase over the next two to five years and 45% thought that it would increase over the longer period of five to ten years. The energy and transport sectors, the larger firms and existing SRCCS users were much more confident about the increasing requirements for SRCCS. Very few sites (only five) considered that there would be a reduced requirement and all were SRCCS users. Some 35% were unsure about the direction of change in either the medium or long term nearly all of whom were currently not using SRCCS.

111 The expectations of the sample with regard to increased requirements for SRCCS were matched closely by anticipated increases in expenditure on such systems. The main reasons quoted for increased expenditure were that new technology would become available (including increased use of computers in manufacturing) and that new legislation or requirements would be introduced.

112 Nearly 30% of the sample considered that expenditure on SRCCS would stay much the same over the next two to five years. The main reason for this was that SRCCS was not thought to be required or appropriate for the particular industry or that investment in the sector would not be sufficient to warrant the introduction of SRCCS - a sentiment expressed most strongly in discrete and process manufacturing. A less often quoted reason was that the cost-benefit calculation could not justify installation of the systems - this was quoted most often in the energy sector.

113 Although one third or more of the sample anticipated increased requirements for, and increased expenditure on, SRCCS over the medium term, only 10% (50 sites) planned to implement such systems in the next year. Planned implementation was much higher in the energy and transport sectors and was dominated by existing SRCCS users.

114 Table 4 summarises the survey results on expected future use of SRCCS by sector and site size.

Table 4: Expected increased use of SRCCS (% all sites)

	Total	Discrete	Process	Energy	Transport	Small	Medium	Large
		Manufacture				Site Size		
Requirement increase 2-5 yrs	35	33	32	45	66	23	35	44
5-10 yrs	45	43	42	43	79	36	45	48
Increased expenditure 2-5 yrs	35	30	32	43	69	20	35	44
Planned implementation 12 months	10	5	10	29	21	3	11	12

115 Table 5 presents the expectations about increased SRCCS requirements, expenditure and implementation by type of system currently in place. It suggests that SRCCS users and particularly specialist SRCCS users are likely to be the main source of growth in the market. There is little evidence of a significant shift from non-use to use of SRCCS.

Table 5: Expected increased use of SRCCS by type of current use (% all sites)

	Current SRCCS users			Non - SRCCS Users
	Total	Specialist	Other	
Requirement increase				
2 - 5 yrs	60	70	50	26
5 - 10 yrs	70	72	69	35
Increased expenditure				
2 - 5 yrs	59	70	53	25
Planned implementation 12 months	29	36	25	3

116 There were 50 sites planning to implement an SRCCS in the next 12 months. They were distributed across the range of industrial sectors but with some concentration in energy and transport applications. There were no particular types of system likely to be favoured in future implementation plans, nor was there any discernible preference for package or bespoke systems.

117 The planned systems seemed intended to cover either frequent risk (ie many times per annum) or remote and improbable risk. There is some correlation with the sector of intended application as Table 6 shows. Remote/incredible risk is the concern of the process and energy sectors and frequent risk that of the transport sector.

118 Over 60% of those planning to implement a system in the next 12 months were doing so to prevent malfunction of, or damage to, the process system or plant; between 50% and 60% were doing so to prevent injury to the workforce and operators; and 30-40% aimed to prevent environmental damage or injury to the general public.

Table 6: Frequency of risk and sector of application for planned systems (number of sites)

	Total	Discrete Manufacture	Process	Energy	Transport
Frequent risk (at least once per annum)	13	-	5	2	6
Occasional	2	1	-	1	-
Remote and incredible risk	20	3	8	9	-
Don't Know	15	6	7	2	-

119 About 45% of those planning procurement of SRCCS in the next year did not know how much investment was likely to be involved. The remainder provide an inadequate sample on which to base any general conclusions. Six sites were planning to spend between £1-27 million each on their system, all but one of these sites being in energy or process manufacturing. Seventeen of the sites were planning expenditure of less than £½ million, 14 of these being in manufacturing.

120 There does not appear to be a strong information base on which the sample could base its judgements of future safety control requirements and appropriate systems. Trade journals were quoted as the most useful sources of information - but then only by 17% of the sample. Even amongst users of SRCCS, only 24% quoted trade journals and 13% equipment suppliers as useful information sources. Over 40% of the sample claimed no knowledge about the relative usefulness of such sources.

121 Only about one quarter of the sample considered that information was readily available - about 35% said it was difficult to find or not available at all and 37% claimed they didn't know whether it was or not. Thirty five per cent claimed more information was needed to take advantage of the benefits of safety control systems. This represented 175 sites of which over one half claimed that they needed general information on everything - there was no significant distinction between types of system use and site size bands.

Summary - all sites

122 The picture painted by the survey of all sites using safety control systems is one of both distinctions and similarities. The main distinctions are that:

- the energy and transport sectors and the larger sites were characterised by higher penetration of use of SRCCS, by awareness of SRCCS related standards, and by higher expectations about increased requirements for, expenditure on and implementation of SRCCS over the short to medium term;

- penetration of SRCCS use in manufacturing (both discrete and process) and in small firms was lower than that of other safety control systems;

- current users of SRCCS, particularly the specialist users, were much more likely to anticipate increased requirements for, expenditure on, and implementation of SRCCS over the short to medium term than users of other safety control systems.

123 The similarities amongst the sample subsets were primarily to do with the lack of awareness and information:

- in no subset (even in the energy and transport sectors and amongst SRCCS users) did the level of awareness of SRCCS related standards exceed 65% and in most cases it was 50% or less - amongst non-SRCCS users' awareness was virtually zero;

- in most subsets, the proportion claiming that information was difficult to find or not available was one third or more, even amongst specialist SRCCS users (the only exception was the transport sector);

- more information was thought desirable by over one third in each subset except amongst small firms (where the responses were dominated by "don't knows").

Sites using safety control other than SRCCS

124 There were over 350 sites in the sample classified as non-users of SRCCS. These were primarily in manufacturing (95%) and small/medium sized sites (77%). They used a mixture of safety control systems (and more than one on each site). A significant proportion was not certain about the control system in use. Table 7 summarises the sample in these terms.

Table 7: Non-users of SRCCS - sample characteristics

	Total	Discrete Manufacture	Process	Energy	Transport	Small	Medium	Large
						Site Size		
Number of sites	356	173	165	8	10	106	168	78
% using:								
• hardwire	48	46	53	25	30	44	49	55
• mechanical	46	43	52	12	40	36	51	53
• other	44	44	46	12	50	41	46	47
• none	1	1	2	-	10	-	1	4
• not sure	31	34	26	62	30	35	29	27

125 About 30% of the non-users were aware of at least some SRCCS systems. More than 40% were unaware and the rest were uncertain. This suggests that less than one third of the sample was aware of the well-defined SRCCS "product". This proportion was markedly lower amongst the smallest sites.

126 There were some 100 sites aware of SRCCS systems. Only one quarter had considered adopting them. Table 8 summarises the main reasons either for not considering the SRCCS option or for rejecting it.

Table 8: Reasons for not considering or rejecting SRCCS by those aware of them (% sites)

	Not Considered SRCCS	Rejected SRCCS
Reason • existing technology sufficient	49	35
• process not dangerous enough	18	4
• no cost-benefit from implementation	18	35
Number of sites	76	26

127 The lack of consideration of SRCCS was primarily because the existing technology was regarded as sufficient given the process dangers and the cost-benefit ratio from implementation. Rejection of SRCCS after consideration clearly was not attributable to low levels of perceived process danger but, for one third of those who had rejected SRCCS, was based on an assessment of the cost-benefit ratio of implementation relative to existing technology.

128 Amongst those who were aware of SRCCS, there was an even split between those who considered them to have disadvantages and those who did not. The larger sites tended to assess SRCCS less favourably, possibly because they had more experience and knowledge of SRCCS and their requirements and limitations. There were two distinct types of disadvantage associated with SRCCS:

- the need for the systems to be fail-safe and their lack of reliability in these terms;
- the cost and resource intensity of such systems, particularly their requirements in terms of technology and programming.

129 Only 8% of the 356 non-users was aware of SRCCS related standards but one quarter thought that requirements for, and expenditure on, SRCCS would increase over the next two to five years. About one quarter considered that the key issue over the medium term would be a significant increase in regulations and standards. The consequence of this was thought likely to be increased expenditure on SRCCS over the next two to five years. However, some 25% said that expenditure would stay the same,

the overwhelming reason being that SRCCS was not required or appropriate for the respondent's sector - mostly in manufacturing.

130 Over the short-term (the next 12 months) very few sites amongst current non-users (about 3%) were intending to implement SRCCS. There was no clear pattern to the limited number of sites planning to instal SRCCS. All that can be said is that most were planning to introduce only one system either in package form or a combination of package and bespoke, directed at guarding against loss of life with a frequency of risk which was either regarded to be very high (ie many times a year) or very low (ie remote or improbable).

131 A significant minority (over 40%) of non-users claimed not to know whether and how use of SRCCS should be encouraged. Of those that had views the main factor likely to encourage use was a combination of increased cost-effectiveness (18%) and improved reliability (10%).

132 Other action which could be directed at improving safety control might include training and information provision. Nearly two-thirds of non-SRCCS users provided some training related to the safety of the process - this proportion was much the same for the small sites as for the large. Much of the training was in-house (54% of sites) and involved induction courses (49%). But, beyond that, training was very limited. There was no overwhelming demand for more formal training in the safety area - one quarter said there was no need and over 30% said they were not sure. The rest - some 45% representing about 160 sites - clearly provides a potential demand for more formal training which might be linked with the very low proportions of non-users providing training with regard to system specification, design, use and maintenance - see Table 9.

Table 9: Training practice and need (% non SRCCS users)

	Total	Discrete Manufacture	Process	Energy & Transport	Small	Medium	Large
					Sites		
Safety related training provided	64	60	69	56	57	68	65
Type of training*							
In-house	54	57	52	50	65	53	45
Induction	49	41	55	60	27	59	53
System use	14	16	11	30	17	10	18
Other system attributes	6	6	4	30	7	3	12
More formal training on safety required	44	42	48	28	38	49	45

* As a percentage of the total sites providing training

133 The sources of information used by non-users were wide-ranging but few respondents claimed any one of them to be particularly useful. For example, about half the non-users claimed to use trade journals but only 17% claimed them as the "most" useful source of information.

134 Only 17% of non users claimed that information was readily available on safety related control systems - about one third thought it was difficult to find and half classified themselves as "don't knows". About one quarter considered that no more information was needed to take advantage of the benefits of SRCCS, although one third did and over one half of these wanted general information on everything. Nearly 20% (65 sites) expressed an interest in joining the SRCCS association or club.

Summary - sites using safety control other than SRCCS

135 From the survey of the sample of non-users of SRCCS it would appear that this group is not likely to be a major source of new demand for SRCCS even over the medium term. There are at least two reasons for this suggested by the survey:

- There was a proportion of non-users (about one quarter to one third) who were aware and alert to the potential for SRCCS and the likely growth in requirements or standards based on such systems. These non-users had examined SRCCS and, for reasons largely to do with system reliability and resource intensity of system management, had decided against adoption of the systems. Improvements in technology and increases in regulation would see this group re-examining the options in a relatively informed way.

- There was a larger proportion of non-users (about two-thirds to three-quarters) who appeared either to be indifferent or ill-informed about SRCCS and its developments. It is not clear from the survey whether any more information would change attitudes although there is a suggestion that more formal training and better quality information would be welcomed by some part of this group.

136 The first group of non-users was distinguished from the second in terms of site size, eg only 18% of small sites was aware of any SRCCS whilst 38% of large firms claimed such awareness. The energy sector was well represented in the first group but it was not obvious that transport and discrete or process manufacturing characterised the first any more than they do the second.

Sites using SRCCS

137 There were 141 sites in the sample which had implemented SRCCS. The sample was evenly distributed across discrete manufacturing, process manufacturing and the energy sector. But only about 10% were sites employing 100 or fewer.

138 The results of the survey of SRCCS users were organised in order to provide insights into a market distinction which had emerged from the supply-side investigation as being potentially significant. The distinction was between specialist users of SRCCS and other users. The former used SRCCS specifically to control for safety (rather than any other control purpose) and relied on specialised suppliers of such equipment. On the other hand, other users of SRCCS tended to use the equipment for safety and other control purposes. There were a number of hypotheses or expectations about these two groups which had arisen elsewhere in the study and which the demand-side survey could help to test:

- the specialist users of SRCCS represented the demand-side of a more well-defined market - others users would be more difficult to define as a distinct market at all;

- the specialist SRCCS market was likely to be more mature and confined to specific applications;

- the greater maturity of the specialist market could be expected to be reflected in more well-developed relationships with suppliers, understanding and use of guidelines and standards for system design, specification and validation, and training.

139 In order to test these propositions and to identify any other distinctions between these two groups of users, the sample was organised for reporting purposes to separate out users - termed specialist users - who identified their suppliers as those generally regarded from the supply-side as being suppliers of specialist safety-related computer controller equipment and systems. This selection was validated by examining the extent to which the specialist suppliers claimed to be users of dedicated equipment - a self-classification related to the degree to which the primary purpose of the system used was for safety control.

Table 10: Specialist and dedicated users of SRCCS

	(Number of Users)		
	Dedicated	Non-dedicated	Total
Specialist users	35	15	50
Other users	35	56	91
Total	70	71	141

140 Table 10 summarises the results of this validation exercise. 30% of the specialist users claim to be using non-dedicated systems. This suggests that the specialist user group was not uniquely defined by this method of organising the sample. But, it should be noted that just over 60% of "other users" claimed to be using non-dedicated systems. There does therefore appear to be a close relationship between specialist and dedicated users. The rest of the report on the survey persists with the distinction between specialist users and others as a more accurate depiction of the market distinction than between dedicated and non-dedicated users.

141 The sample of SRCCS users is more fully described in Table 11. It demonstrates that specialist users were more prevalent in the process and energy sectors and especially in transport than other users and very much less represented on the small sites and in discrete manufacturing.

Table 11: Users of SRCCS - sample characteristics

	Total	Discrete Manu-facture	Process	Energy	Transport	Small	Medium	Large
						Sites		
Number of Sites								
• all users	141	39	41	41	19	16	60	55
• specialist users	50	4	15	16	15	3	18	22
Mean Number of Systems								
• all users	16	11	12	6	56	4	7	22
• specialist users	25	12	9	11	60	2	8	32

142 Expenditure on the systems varied from less than £50,000 to more than £45 million. About 45% of the users had invested less than £½ million each but ten sites had invested more than £2 million mostly in the energy and process manufacturing sectors. About one third of the sample either did not know, or refused to divulge, their investment expenditure. This limits the conclusions that can be drawn from the survey on the question of system costs. However, it would appear that the systems in use in the process and energy sectors were very much more expensive (in excess of £2 million) than in discrete manufacturing and transport (between £500,000 and £800,000).

143 Table 12 presents the evidence from the survey on the build-up over time of systems installed - both by number of systems and number of users with systems.

144 There was an increase in the rate of installations of SRCCS both by system and by user during the latter part of the 1980s and the early 1990s. The proportion of installations represented by specialist systems declined from 55% in 1980 to 40% in 1991. Figure 1 presents this data in graphical form which more clearly demonstrates the more rapid growth in non-specialist systems installed. Figure 2 presents the total number of installations broken down by industry sector.

Table 12: Cumulative sum of systems installed and of users with systems installed

	Number of Systems		Number of Users	
	All Users	Specialists	All Users	Specialists
1980	36	20	26	14
1985	75	39	51	22
1986	88	46	63	28
1987	103	52	77	33
1988	123	57	91	37
1989	153	69	109	46
1990	177	76	130	53
1991	215	86	162	62
Don't know	56	37	13	5

Note : These figures are the cumulative sum of systems installed. They do not necessarily represent the stock of systems in use since no allowance has been made for scrapping.

145 The distribution of system installations over the last ten years and by sector and site size is presented in Table 13. The maturity of the energy and transport markets for SRCCS is clearly demonstrated - the number of installations in fact declined over the 1980s in the transport sector. The specialist systems were of particular importance in those two markets but showed no growth in installations over the period.

146 The infancy of the market in the early 1980s for SRCCS in discrete and process manufacturing is shown by the number of installations in Table 13. But the rapid growth of the market is also very clear from the data. In the process sector this was associated with a growth in the use of specialist systems such that in the period 1989-1991 more were installed in this sector than in all others in the sample.

Figure 1 : Cumulative Number of SRCC Systems Installed; Survey Results

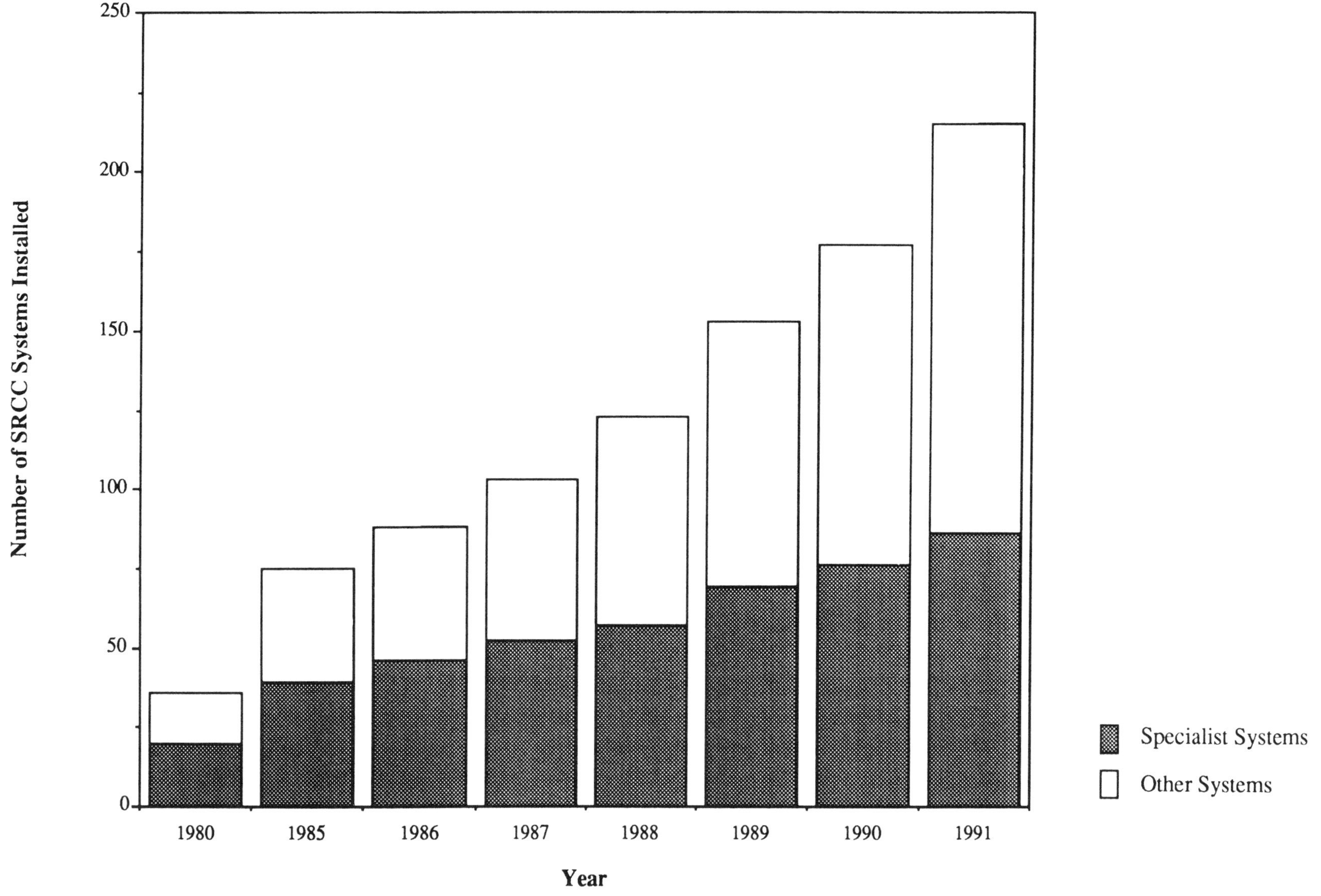

GSE-32-16

Figure 2 : Cumulative Number of Sites Installing SRCC Systems; Survey Results

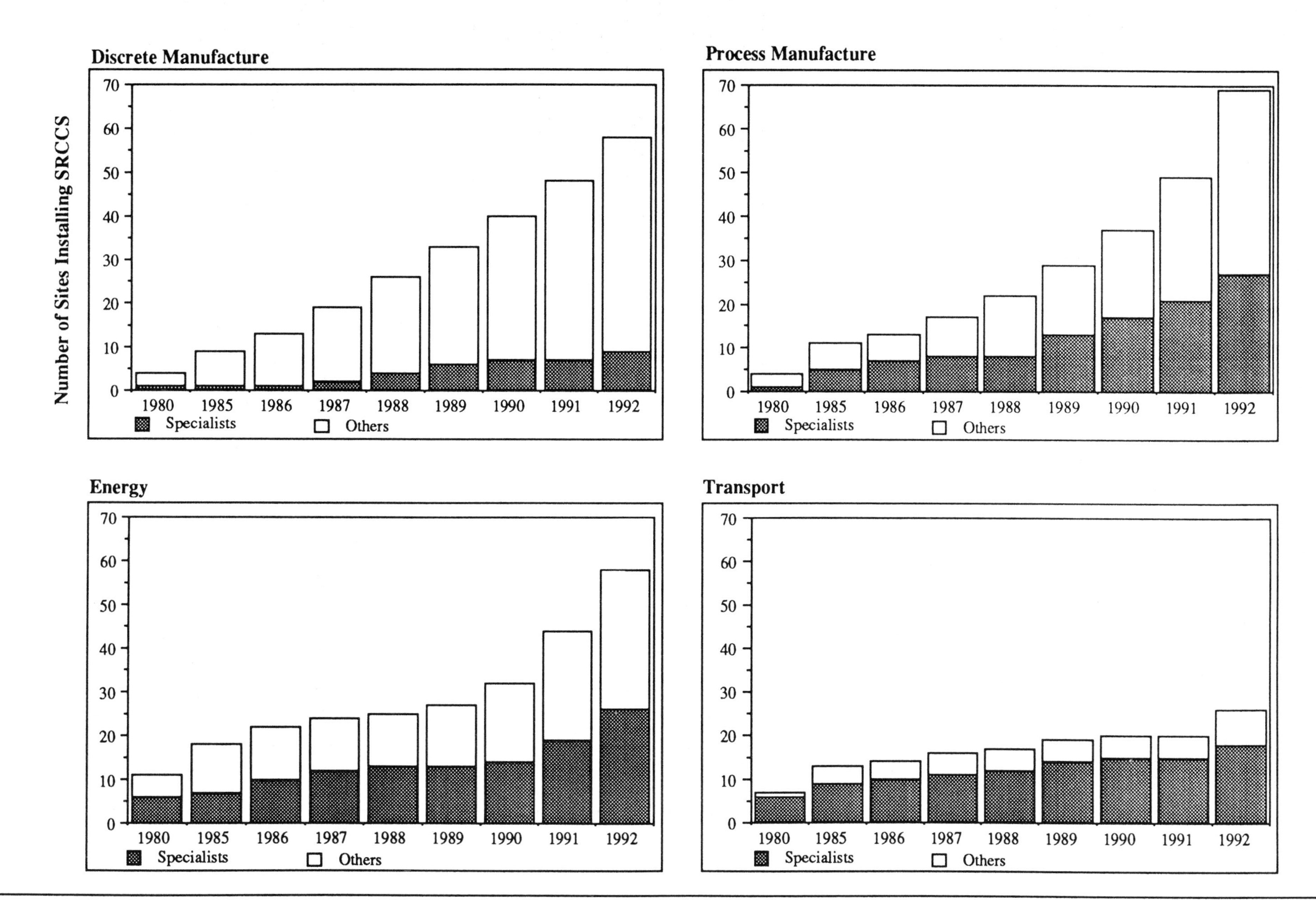

GSE-32-17

147 The use of SRCCS grew rapidly over the 1980s in medium and large sites. It was only at the turn of the decade that the small sites increased their use of SRCCS significantly and the take-up of specialist systems remained very low amongst these sites.

Table 13: Distribution of new systems installed

	Total	Number of New Installations						
		Discrete Manufacture	Process	Energy	Transport	Small	Medium	Large
						Sites		
Pre 1980	26(14)	4(1)	4 (1)	11(6)	7 (6)	3(1)	9(3)	10 (6)
1980-1985	25 (8)	5(-)	7 (4)	7(1)	6 (3)	2(-)	14(3)	7 (3)
1986-1988	40(15)	17(3)	11 (3)	7(6)	4 (3)	4(-)	17(4)	17(10)
1989-1991	71(25)	22(3)	27(13)	19(6)	3 (3)	10(2)	37(13)	23 (8)

Note: Figures in brackets are numbers of specialist system installations.

Table 14: Frequency of hazard (% users)

Frequency	Total	Discrete Manufacturing	Process	Energy	Transport	Small	Medium Sites	Large
• At least once a year	47(58)	34(25)	32(40)	56(56)	89(87)	38(33)	36(44)	56(64)
• Occasional	14(10	26(25)	12(13)	12(13)	-(-)	19(33)	22(11)	7(9)
• Remote incredible	38(34)	36(50)	52(53)	39(38)	5(7)	51(66)	43(45)	30(32)
• Don't know	6(6)	5(-)	10(7)	5(6)	5(7)	-(-)	5(6)	9(5)

Note: % of users will not sum to 100 because each site may have more than one system covering different frequencies of hazard.

Figures in brackets refer to the % of specialist systems.

148 Table 14 presents the broad distinction between users in terms of the frequency of hazard being protected against. Bearing in mind the smallness of the sample for specialist system users (at least for the purposes of the detailed disaggregation shown in Table 14), there would appear to be no evidence that specialist systems were being used to protect against either higher or lower frequency of hazard occurring. The transport sector used SRCCS to protect against high frequency of hazard but there was no clear pattern in the other sectors except perhaps a hint that the process sector sought to protect against remote/incredible hazard to a greater extent than other sectors. A general observation from Table 14 is that in all sectors the preoccupation appeared to be with either high or low frequency of hazard rather than occasional occurrence of hazard (ie during the product/process lifetime).

149 Table 15 presents the nature of the hazard which different sectors and site sizes were protecting against by use of SRCCS. 70-80% of users were concerned to protect against the hazard of malfunction of the process and damage to plant; 50-70% were concerned about protection against injury to the operators and general workforce; and 25-50% were seeking to prevent environmental damage, spillage of dangerous substances and injury to the general public. There is little evidence from these data that specialist SRCCS users had adopted such systems to protect against hazards in a pattern significantly different from other SRCCS users with the possible exception of process manufacturing.

Table 15: Nature of hazard (% of users)

	Total	Discrete Manufacture	Process	Energy	Transport
Process malfunction	76 (82)	64	63 (67)	90 (94)	95 (100)
Plant damage	72 (72)	64	76 (87)	88 (81)	47 (47)
Operator fatal injury	67 (58)	90	85 (93)	39 (38)	42 (40)
Workforce fatal injury	61 (52)	59	73 (67)	56 (44)	47 (40)
Non operator fatal injury	52 (50)	59	61 (60)	39 (31)	53 (53)
Environmental damage	46 (48)	26	46 (60)	78 (81)	16 (7)
Spillage of dangerous substances	39 (36)	31	41 (53)	61 (56)	- (-)
General public injury	28 (38)	13	15 (13)	27 (25)	89 (87)

Note: Figures in brackets refer to specialist system users - no figures for discrete manufacturing are given because of small sample size.

150 Process malfunction, plant damage and environmental damage were of particular concern in the energy sector. Process malfunction and injury to the general public were the concerns of the transport sector. Operator fatal injury was the main preoccupation in manufacturing. These different concerns are clearly a reflection of the nature of the process and the business at each of the sites in the sample.

151 Taking frequency of risk and the nature of the hazard together the following summary picture can be drawn of the pattern of SRCCS use:

- the manufacturing sector adopted SRCCS primarily to cope with both relatively low and high frequency risks liable to lead to fatal operator injury - process manufacturing was somewhat more concerned with lower frequency risk liable to lead to environmental damage;

- the energy sector's interest in SRCCS was to cope with both low and high frequency of risk with somewhat greater emphasis on the latter and where process malfunction, plant damage and environmental damage were the primary hazards of concern;

- the transport sector adopted SRCCS to deal with high frequency risks (many times a year) of process malfunction and general public injuries.

152 This pattern of control for safety using SRCCS is brought to life by consideration of the major applications in each of the broad sectors (see Table 16). There was a broad spread to the applications, the primary ones being:

- machinery such as milling and trimming;
- iron and steel casting;
- power generation;
- water supply;
- coal mining;
- air traffic control;
- road and rail traffic control.

153 There were few applications in which there were three or more specialist SRCCS users. These were:

- air, road and rail traffic control;
- power generation;
- water supply;
- petrochemicals;
- packing/packaging/paper.

Table 16: Specific applications of SRCCS

Sector (Number of Sites)	Applications in 10%+ of Users	% of Users in Each Sector
Discrete manufacture (39)	• Machinery (eg milling and trimming) • Iron and steel (eg casting)	23 26
Process manufacture (41)	• Machinery (eg milling etc) • Heat treatment (eg baking) • Rubber product processing • Packing/packaging/paper • Textile processing • General incl. food processing	15 12 10 15 * 12 10
Energy (41)	• Power generation - monitoring and control of substations • Water supply - pollution overflow • Coal mining/environment underground • Heat treatment and processing • Power generation - fuel management and production	22 * 20 * 20 12 15 *
Transport (19)	• Air traffic control • Traffic control - road and rail • Ports/shipping	32 * 32 * 16
Site Size (Number)	**Applications**	**% of Users in Each Size Band**
Small (100 or fewer employees) (16)	• Machinery • Iron and steel • Water supply	31 19 13
Medium (101 - 500) (60)	• Iron and steel • Rubber products • Packing/packaging/paper	12 10 10 *
Large (over 500) (55)	• Machinery • Power generation • Petrochemicals	13 13 * 11 *

* Sectors and site sizes in which there were three or more specialist SRCCS users.

154 The protective features of SRCCS most commonly used in each sector were geared to the particular hazards with which ease sector must cope. Table 17 identifies these features.

Table 17: Protective features most commonly used in each sector

	% Users				
	Discrete manufacturer	Process manufacturer	Energy	Transport	Total
• Automatic cut-off or shutdown	**41**	**37**	29	-	30
• Voting system	28	27	32	16	27
• Alarms/warning/ hazard lights	21	12	**51**	16	26
• Monitor of mix of ingredients	3	7	7	**63**	13

155 The decision to opt for some kind of SRCCS appeared to be driven primarily by the imperative to minimise the risk (because of the nature of the process or the degree of hazard) and to comply with regulations and standards. All the sectors of use agreed that these factors were of primary importance (see Table 18).

Table 18: Importance of factors in selecting SRCCS (% users)

	Total	Discrete Manufacture	Process Manufacture	Energy	Transport
First-order factors					
• need to minimise risk	99	100	98	98	100
• process demands safety	89	87	98	85	79
• compliance with HSE	89	97	95	85	63
• site adheres to standards	85	77	85	90	89
• degree of hazard	75	77	83	71	68
Second-order factors					
• cost-benefit analysis	68	67	66	66	79
• in-built in process	65	85	76	54	26
• used in similar sectors	57	67	56	51	53
Third order factors					
• cost of training	45	49	49	41	42
• company reputation	44	44	46	41	47
• third party recommendation	36	33	27	41	47

156 There was a second-order set of factors and here some clear differences emerged between the sectors. In particular, for discrete and process manufacture it was important that the SRCCS was built into the process machinery. This was less important in the energy sector and regarded to be of little importance in transport.

157 A third order set of factors (less than half the users considered these to be important) included the cost of training involved in system use and the enhancement to the company's reputation brought about by use of the system.

158 In general, there were no significant systematic differences in the factors for selecting SRCCS between the specialist users and others. One difference worth highlighting was that the specialists were not quite so insistent that the SRCCS should be built into the process control system - 50% of all specialist users compared with 75% of other users. There is some evidence from this to suggest that the specialist SRCCS user was indeed primarily concerned with the safety purpose of the system - even in discrete manufacturing where most users wanted the safety system integrated with the control system.

159 Once a decision had been made to opt for SRCCS, there were a number of critical factors which were considered in the acquisition of a specific system. The survey respondents in all sectors and site size bands were agreed on these factors but it was not possible to distinguish between each of them in terms of their relative importance (except that price and ease of installation of the system were clearly lower in the rank order). In Table 19 the factors have been put into three bands representing the ranges of sample proportions who thought the factors were important.

Table 19: Important factors in acquisition of a SRCC system

Factors rated as important by over 90% of all SRCCS users	% of users
• ease of use	96
• functionality	93
• ability of system to meet exact needs	93
• level of integrity	92
• fault tolerance levels	91
• level of maintenance	91
Factors rated as important by 80-90% of all SRCCS users	**% of users**
• reputation of the supplier	89
• product guarantee	87
• level of external support	84
• cost of maintenance	81
Factors rated as important by 60-80% of all SRCCS users	**% of users**
• cost-effectiveness	79
• price	63
• ease of installation	63
Note : The transport sector was the one most likely to vary from the overall picture - eg only 60% of users in this sector considered costs of maintenance important and only half were concerned about ease of installation.	

160 There were no systematic differences between specialist and other users in the factors determining the acquisition of a particular SRCCS. Supplier reputation and levels

of external support were important to somewhat higher proportions of specialist users and product guarantee to somewhat lower proportions.

161 In deciding to opt for SRCCS and to acquire a particular system there are a number of steps in implementation. The specifier of the type of system to be implemented was supplied from in-house expertise in the overwhelming majority of SRCCS users (nearly 90%). Only one third used specific techniques to specify their requirements (just over 40% in the case of specialist SRCCS users). Hazard analysis had been used on only two sites - in the energy sector. The most popular technique was said to be "in-house experience" (but only 16% of those using techniques claimed this). However some 40% of those who claimed to use specific techniques could not say what they were.

162 Specification guidelines for system design were used by the majority of SRCCS users (70%) of which again the most popular were either in-house guidelines (but only 17% of those using guidelines) or in-house reports on the work done (15%). External standards of two types were given specific mention - HSE/IEE guidelines (16%) and relevant British Standards (10%).

163 "In-house experience" for specifying system requirements and in-house guidelines or in-house reports for system design were slightly more popular amongst the specialist users than other users of SRCCS. Table 20 summarises the findings of the survey on the specification of system requirements and design.

164 The "do it yourself" approach to system specification and design suggested by the above did not present any difficulties for about 30% of the users. But for others, particular problems were encountered with:

- the timescales involved in design (32% of users);
- difficulties in design (22%);
- lack of in-house design skills (21%);
- cost of designing the system (18%).

Table 20: Specification of SRCCS requirements and design

	(% of users)				
	Total	Discrete Manufacture	Process Manufacture	Energy	Transport
Specifier of system • in-house experts	88 (88)	77 (75)	88 (93)	93 (94)	84 (80)
Techniques used to specify system requirement - yes	35 (42)	36 (50)	32 (33)	32 (38)	42 (53)
Guidelines used to design system - yes	70 (70)	54 (75)	71 (60)	83 (75)	68 (73)
Of which:					
• in-house	32 (43)	38 (33)	40 (78)	27 (34)	23 (27)
• external standards	26 (25)	29 (-)	20 (22)	24 (16)	46 (45)
• don't know	18 (23)	19 (33)	10 (11)	18 (25)	31 (27)

Note: Figures in brackets refer to specialist SRCCS users.

165 The time taken between system specification and operational status varied for different sectors according to the size and complexity of the systems typically associated with each sector (see Table 21). Most discrete manufacturing SRCCS were operational within three months; typically, process manufacturing systems took a little longer; energy and transport sector systems could take longer than six months or even one year to become operational. However, there is little evidence that these schedules were any longer than those that had been anticipated by users (see Table 22).

Table 21: Time taken for SRCCS to become fully operational

	Total	(% Users)			
		Discrete Manufacture	Process Manufacture	Energy	Transport
Less than 1 month	25	46	27	7	11
1 month - 3 months	17	15	29	10	11
3 months - 6 months	15	5	17	27	5
6 months - 12 months	11	13	5	15	11
12 months plus	16	3	10	29	26
Don't know	17	18	12	12	37

Table 22: Actual compared with expected schedule to fully operational SRCCS

Expected Time Taken	Actual Time Taken				
	Less Than 1 Month	1 to 3 Months	3 to 6 Months	6 to 12 Months	12 Months or Longer
Less than 1 month	32	-	-	-	-
1 - 3 months	-	17	-	-	-
3 - 6 months	-	-	19	7	-
6 - 12 months	-	-	-	6	-
12 months or longer	-	-	-	-	15

166 Once the system was in place, its validation was undertaken in-house by 40% of users (lower in discrete manufacture and higher in the energy sector and for specialist SRCCS users) and using specific guidelines on some 80 sites (ie 56% of the sample). Of the latter about one quarter used in-house guidelines, another quarter used "final quality instrumentation" and, amongst the range of other techniques or guidelines, most appeared to be of a physical or mechanical kind rather than based on software. The process of validation could take a long time - in 17% of users and for 35% of users in the energy sector it took more than three months.

167 The maintenance of the systems was carried out by in-house experts in over 80% of users but about 45% also used the original system supplier (a higher proportion - some 70% - in the transport sector). Most users claimed that their maintenance was either "very satisfactory" (55%) or "fairly satisfactory" (35%).

168 It will be apparent from the survey results with regard to procedures for system acquisition, installation, validation and maintenance that a significant proportion of users relied on in-house expertise. This "do-it-yourself" approach was slightly more pronounced amongst the specialist SRCCS users. This suggests that training ought to be an important feature in user companies. Table 23 presents the proportion of users who had adopted training for different aspects of SRCCS.

Table 23: Training provided by users of SRCCS (% of users)

	Total	Discrete	Process	Energy	Transport	Small	Medium	Large
		Manufacturing				Sites		
System								
• use	89	92	80	90	100	88	90	89
• maintenance	77	74	66	83	89	56	82	76
Hazard assessment	69	74	68	59	84	63	70	71
System								
• installation	51	49	49	56	47	19	55	58
• validation	50	46	44	61	47	38	50	53
• specification	42	41	34	49	42	25	35	56
• design	33	28	39	34	26	38	30	36
No training	10	8	17	10	-	6	10	11

169 Training in system use and maintenance and in hazard assessment was provided in a very high proportion of users especially in the transport sector. The proportion providing training declined in all sectors the more the training was focused on the earlier stages of system design and specification. This suggests that the extent of reliance on in-house expertise for specification of the systems (see Table 20) may not be supported by adequate training in system specification and design. In particular, it is worth noting the high proportion of users who claimed to provide training in hazard assessment and the contrast with the very few users who had adopted hazard analysis to assess system requirements. The smaller sites were less likely to undertake training, although nearly 90% of them provided training in system use.

170 There was little reliance amongst users on external sources of training alone (only 17% of users). The evidence of the survey suggests therefore that the "do-it-yourself" approach to system specification and design extended to the training needed to support this function.

171 There is a clear contrast between users and non-users of SRCCS in terms of training provision as revealed in the comparison of Tables 9 and 23. Only 14% of non-users provided training in safety system use compared with nearly 90% amongst users.

172 Although there was a very high proportion of SRCCS users which provided training in system use and maintenance, the commitment to training in these companies was significant only for a relatively low proportion of companies, there was little knowledge about the commitment to training, and maintenance and operator staff were the main beneficiaries of training (Table 24).

173 There was general consensus that training was "very useful" (some 75% of users and over 90% of small sites). Despite this widely held appreciation of training, there was less enthusiasm for more formal training in use of SRCCS. Only 43% claimed this; about one third of the latter wanted more formal training to keep up with technical advances and some 20% because of a lack of understanding of the system. The specialist SRCCS users who wanted more formal training had a very different view - two-thirds wanted to keep up with the technology and just 9% were concerned about their limited understanding of the systems. For the 80 sites which did not want further formal training, nearly 40% claimed that there was enough training already and some 30% said that "safety was already built into the system" or that the danger inherent in the industry had already established the need for training. This evidence suggests that the "do-it-yourself" approach was accompanied by complacency amongst a significant proportion of SRCCS users.

Table 24: Commitment to training by users of SRCCS (% of users)

Person days given up for training	
1 - 15	35 (20)
15 - 30	13 (11)
31 - 100	12 (16)
101 - 1 year	9 (11)
Don't know	28 (38)
Investment in training	
Less than £15,000	13 (16)
£15,000 - 250,000	7 (4)
£250,000 - 1 million	2 (2)
Don't know	74 (71)
Job title attending training	
Engineers (maintenance)	46 (53)
Operators/machinists	46 (42)

Note: Figures in brackets refer to specialist SRCCS users.

174 The commitment to SRCCS amongst users was clearly revealed by the 30% of users who intended to implement further systems in the next year - about 40 user sites. Taking account of the limited number of non-SRCCS users who intended to shift to SRCCS, this represented a total of 63 systems planned for the next 12 months. If these plans come to fruition, the build up of the number of systems installed (not the stock of systems in use because no allowance can be made for scrapping) will be as set out in Table 25.

175 The intentions revealed by the survey indicated a renewal of growth in the installations of SRCCS in the transport and energy sectors and continued or faster growth

in process and discrete manufacturing. As a consequence there were no discernible differences between the features of the planned and the existing stock of SRCCS in terms of the frequency and nature of the hazard which the systems were designed to protect against.

Table 25: Cumulative number of systems installed

	Number of Systems Installed
1980	36
1985	75
1990	177
1992 (potential)	278

176 SRCCS users and particularly specialist users were much more likely than non-users to expect increased requirements for, and expenditure on, SRCCS over the medium term. But the reasons were the same - technology advances and new legislation or regulatory standards. How did they learn about these? In much the same inadequate ways as non-users it would seem from the survey. The information sources on new developments deemed most useful by users were trade journals and equipment suppliers, but only 24% and 13% of users respectively thought this. About 40% of users (a slightly higher proportion of specialist users) claimed that information was difficult to find and recommended that more information was needed. For those who wanted more information, some 45% just wanted general information on everything - this was the case for nearly two thirds of specialist SRCCS users who thought more information would be useful.

177 The relatively high demand for more information of a general kind was indicative of a lack of awareness of the issues raised by SRCCS and was reinforced by the proportion of respondents who expressed an interest in joining the safety related computer controlled systems association or club. The survey did not reveal the strength of the interest expressed by respondents but the potential is clear from Table 26. Amongst the non-users of SRCCS in the sample there were nearly 20% who expressed an interest in joining such a club. And, at the other end of the spectrum, nearly half of the sample of specialist users expressed such an interest.

Table 26: Interest in joining the SRCCS association or club

	Number of sites	Proportion of sample (%)
Total sample	131	26
• manufacturing	92	22
• energy	28	57
• transport	11	38
SRCCS users	66	47
Specialist users	24	48
Non-users	65	18

Summary - sites using SRCCS

178 Amongst the SRCCS users there was a distinction between the energy and transport sectors - mature users of SRCCS - and the manufacturing users who generally were much less inclined to adopt SRCCS but were increasingly doing so. The relative maturity of experience in the energy and transport sectors was revealed in some of the operational practices and attitudes described in the survey results.

179 The key distinctions amongst SRCCS users can be summarised as follows:

- The transport and energy sectors were mature users of SRCCS; the manufacturing sector's use of SRCCS had grown very rapidly over the last decade.

- The energy sector used SRCCS to cope with risk of process malfunction, plant damage and environmental damage, this risk being of both high and low frequency; the transport sector had to deal with high frequency risk of process malfunction and injury to the general public; the manufacturing sector's recent increasing interest in SRCCS was directed at risks liable to lead to fatal operator injury.

- The energy and transport SRCC systems tended to be more complex as implied by the length of time taken to make them operational and to validate them. The energy systems were typically much more expensive than either the manufacturing or transport systems.

- Training was important in all sectors, primarily in system use and maintenance and particularly in energy and transport - two-thirds of the sites providing 30 days to one year of training were in the energy and transport sectors.

180 The key similarities amongst SRCCS users (including the specialist users) were that:

- Use of SRCC had been increasingly rapidly, particularly in manufacturing, and was expected to continue over the medium term in all sectors.
- SRCCS was used to protect against both high and low frequency of risk and in ways which increasingly blurred the distinction between safety and other control purposes.
- The decision to opt for SRCCS appeared to be driven primarily by the imperative to minimise the risk and to comply with standards - price and cost-effectiveness were less important in system acquisition choice than the system characteristics (eg ease of use, functionality).
- In all stages of system adoption and use, in-house experts tended to play an important role - the "do-it-yourself" approach.
- The particular problems encountered during system acquisition and installation were to do with design, including the lack of in-house design skills.
- Training provision was a strong attribute of most users but the focus was on system use and maintenance rather than system design and specification, even though the specifiers of the systems were largely in-house.
- Despite the importance attached to in-house expertise, less than half the sample of users wanted further formal training.
- Although SRCCS users appeared to be well-informed and well trained compared with non-users, some 40% claimed that information was difficult to find and suggested that "general information on everything" was what was required.
- A significant proportion of SRCCS users, even specialist users, expressed an interest in joining an association or club of SRCCS users.

II Market characteristics

Introduction

201 The survey of users of safety systems, whether computer controlled or not, was designed to contribute to both the quantitative and qualitative requirements in the terms of reference for the study. This part of the report deals with the use of the demand-side survey to derive quantitative estimates of the market characteristics.

202 The estimates in this section have been calculated by Benchmark Research Ltd using accepted market research procedures for a market having a very small number of high spenders who are not representative of the market as a whole. The mean system values used in the calculations are the means derived from the survey with the following adjustments:

- where appropriate, sites with below 50 employees have been excluded in order to be consistent with data available on the population of sites;
- sites identified as 'atypical' have been treated separately to allow more reliable estimates to be made;
- it is a standard practice to modify the mean values in some cases (amongst typical sites in particular) to reflect factors known about the wider computer control market and thus correct minor sampling biases; this was carried out to produce more reliable estimates of average system values which are lower than those reported in the survey.

203 In order to relate the findings from the survey sample to the universe of sites as a whole it was necessary to calculate a weighting factor for each sector as follows:

For each)	Population =	Weighting
sector)	Sample	Factor

This weighting was used to calculate the number of sites in the UK using Safety Related Computer Controlled Systems (SRCCS) and the value of SRCCs in use.

204 It should be noted that the survey was originally only targeted at users of SRCCS and only after a pause was shifted to cover safety related applications more generally because of the unexpectedly low penetration of SRCCS revealed. This naturally introduced a bias which was accounted for in the translation of sample survey findings to generalisable conclusions. The manufacturing sector was the one which presented most difficulty in these terms.

205 It should also be noted that due to the nature of the population and the survey response received in the transport sector the findings for this sector are less representative than for others because one major user declined to take part in the survey and there being no specific relevant historical data available.

Number of systems installed

206 The average number of systems in use in each sector (with atypical sites removed) is set as follows:

Discrete:	6.3
Process:	6.8
Energy:	3.8
Transport:	14.0

207 These figures are related to the overall penetration of SRCCS to give an estimate of the total number of systems currently in use. These are shown in Table 27.

Table 27: Derivation of estimate of number of systems in use

Sector	Users	Weighting Factor	Average used	Total
Discrete	27	41.4	6.3	6,993
Process	31	37.5	6.8	7,752
Energy	41	18.4	3.82	2,880
Transport	19	14	14.0	3,724
Total				21,349

NB Transport figures exclude CAA sites where information was unobtainable. Since these sites could well have higher than the average number of systems, the figures for transport could be underestimated.

208 Table 28 compares the number of specialist SRCCS in use with other SRCCS:

Table 28: Comparison of specialist and other SRCCS

Sector	Specialist Systems	Other Systems	All systems
Discrete	330	6,663	6,993
Process	1,645	6,107	7,752
Energy	1,123	1,757	2,880
Transport	2,898	826	3,724
Total	5,996	15,353	21,349

209 Specialist SRCCS were only identified at the validation stage of the survey, after the results of focus groups. Therefore the fieldwork questionnaire was not set up to differentiate between specialist and non-specialist systems. The estimates given for

specialist use therefore need to make some assumptions on usage levels compared to non-specialist systems.

Market penetration of SRCCS

210 It was possible to derive a figure for the percentage penetration of SRCCSs in the UK, especially in manufacturing, only after close inspection of the questionnaires completed subsequent to the decision taken in consultation with Coopers & Lybrand Deloitte and the DTI to conduct the balance of the survey on a totally random basis.

211 As the subsequent stage of the sample was totally random and sufficiently large, it was possible to estimate the penetration of the discrete and process manufacturing sectors by relating the percentage of users against the proportion of non-users found:

$$\frac{\text{Non-users}}{\text{Users}} = \text{Penetration}$$

212 This gave the following market penetration estimates for SRCCS:

Discrete Manufacture: 14% (18%)
Process Manufacture: 16% (20%)

213 For the Energy and Transport sectors it was easier to use the sample results to calculate market penetration as these interviews were conducted later in the survey on a purely random basis. The results were as follows:

Energy (including Water): 83% (84%)
Transport: 65% (66%)

214 The figures in brackets presented in paragraphs 303 and 304 are those obtained directly from the overall sample survey. They do not suggest that the sample survey results were unrepresentative.

215 Table 29 summarises the penetration estimates for all sites and specialist SRCCs users.

Table 29: Estimates of SRCCS penetration

Sector	SRCCS Penetration of all Sites %	Specialist SRCCS Penetration of all Sites %
Discrete	14	1½
Process	16	6
Energy	83	32
Transport	65	51

Annual demand for SRCCS

216 Annual demand can be derived in two main ways. The planned installations for the next 12 months can be used to give an indication of the perceived demand for SRCCSs in 1992. It is also possible to examine one particular year as a snapshot of the market. As the survey was conducted partway through 1991 it was not possible to use this year as a measure. Therefore it was necessary to use the most recent complete year for which there were results, in this case 1990.

217 The sequence of calculation of the estimates was as follows:

Number installed in 1990 X Weighting factor = Annual UK Demand (Volume)

UK Demand X Average system cost = Annual UK Demand (Value)

218 The average system costs used are set out in Table 30 below for what have been termed both typical and atypical sites. These figures confirm the need to formulate a method that takes full, but realistic, account of very high spend where it occurs in each sector. This removes skew and provides a more accurate estimation of the characteristics of the market.

Table 30: Average cost per system

Sector	Typical Mean Cost	Atypical Mean Cost
Discrete	£24,895	£6,694,445
Process	£25,093	£1,055,556
Energy	£25,166	£6,694,445
Transport	£99,741	£6,694,445

219 This enables estimates of market demand in 1990 and the expected demand in 1992 to be calculated. These are set out in Table 31 for all sites using SRCCS and for sites using specialist SRCCS.

Table 31: Market demand for SRCCS (£m)

Sector	1990		1992	
	Total	Specialist	Total	Specialist
Discrete manufacturing	80	4	130	6
Process Manufacturing	35	7	80	17
Energy	35	14	50	19
Transport	40	32	40	32
Total	190	57	300	74

APPENDIX G

REVIEW OF EMERGING STANDARDS FOR SAFETY RELATED COMPUTER CONTROLLED SYSTEMS

PREPARED BY

SRD (AEA TECHNOLOGY)

CONTENTS

REVIEW OF EMERGING STANDARDS FOR SAFETY RELATED COMPUTER CONTROLLED SYSTEMS

Any views expressed are those of the author and do not necessarily represent those of the producers of the standards.

1. INTRODUCTION

1.1 General

The main features of the emerging standards for software and safety-related software systems in particular are:

- that software should not be assessed in isolation, ie that the software is just one component of the whole system and a safety analysis of the whole system needs to be carried out

- the incorporation of new tools for assessing the quality and safety of the software

- that hazard analysis should be carried out at the outset and be maintained throughout the lifecycle of the system

A brief resume of some relevant existing standards is given in order to place the emerging standards in context.

1.2 Background Information

The BS5750 suite of standards is equivalent to the ISO 9000 series and the EN29000 series, the only differences being in some of the titles, ie the text is the same except for some of the titles and the reference numbers for the respective standards [Figure 1, page 3].

EQUIVALENT STANDARDS					
BS5750		ISO 9000		EN 29000	
Part	Title	ISO No.	Title	EN No.	Title
1	Quality systems - Specification for design/ development, production, installation and servicing	9001	Quality systems - Model for quality assurance in design / development, production, installation and servicing	29001	Quality Systems - Model for quality assurance in design / development, production, installation and servicing
2	Quality systems - Specification for production and installation	9002	Quality systems - Model for quality assurance in production and installation	29002	Quality systems -Model for quality assurance in production and installation
3	Quality system - Specification for final inspection and test	9003	Quality systems - Model for quality assurance in final inspection and test	29003	Quality systems - Model for quality assurance in final inspection and test
13	Guide to the application of BS 5750 : Part 1 to the development, supply and maintenance of software	9000-3	Guide to the application of ISO 9001 to the development, supply and maintenance of software		

Figure 1: Equivalent standards

2. GLOSSARY

AQAP	Allied Quality Assurance Publications
BS	British Standard
DEFSTAN	Defence Standard
ESA	European Space Agency
IEC	International Electrotechnical Commission
ISO	International Standards Organization
MoD	Ministry of Defence
NATO	North Atlantic Treaty Organisation
PES	Programmable Electronic Systems
RTCA	Radio Technical Commission for Aeronautics (USA)
SCS	Safety Critical System
SRCCS	Safety Related Computer Control System

3. EXISTING STANDARDS

3.1 **BS5750 Part 1:1987 / ISO 9001:1987:** Quality Systems: Specification for design/development, production, installation and servicing

The standard deals with general aspects of quality systems and is regarded as the main standard for establishing a quality assurance framework within an organisation. Many companies are now seeking accreditation to the standard. It is specifically geared to manufacturing industry but guidance documents in the series have appeared which provide advice on its application to specific business areas eg services, software developments (see 4.1 below).

For MoD work, from 1 March 1992, all quality system assessments involving software are to be carried out against ISO 9000 and ISO 9000-3, whether carried out by MoD or a third party.

Compliance with ISO 9000 is specified in IEC 65A (Secretariat) 122 (see section 4.2.2.1 below) as mandatory at all levels of software integrity.

3.2 **ESA BSSC 1A : 1984 :** European Space Agency Software Engineering Standards

These standards are applicable to software in satellites and on ground stations and are of a general nature [Ref 1].

3.3 IEC Standards

3.3.1 **IEC 643 : 1979 :** Application of Digital Computers to Nuclear Reactor Instrumentation and Control

The standard provides a functional classification of computers and their functions in nuclear applications. It is currently in the process of being up-dated.

3.3.2 **IEC 880 :** 1986 : Software for Computers in the Safety Systems of Nuclear Power Stations

The standard is applicable to highly reliable software for computers to be used in the safety system of nuclear power stations, including the safety actuation systems. It is not up-to-date in all details and a supplement is currently being produced (see section 4.2.1 below).

3.4 MoD Standards

3.4.1 **DEFSTAN 00-16 :1984:** Guide to the Achievement of Quality in Software

This standard is a guidance document and identifies quality assurance activities without indicating who should carry them out, ie it provides software quality assurance principles without prescribing the organisational methods to be employed to implement the principles. As with most MoD standards, the standard is based on the procurement lifecycle.

Annex A describes aspects to be considered in the specification of requirements. Annexes B-K provide useful checklists of questions on:

- planning
- design and programming techniques and methods
- documentation
- configuration management
- design reviews
- tests
- trials (acceptance testing) procedure
- transfer to customer
- sub-contracting of software

3.5 NATO Standards

3.5.1 **AQAP 1 : 1985 :** NATO Requirements for an Industrial Quality Control System

This standard was the baseline for quality systems prior to the production of the BS5750/ISO 9000 series and establishes requirements for quality control system elements to be designed, established and maintained by NATO contractors in order to ensure that material and services conform to contract requirements and to provide objective evidence.

3.5.2 **AQAP 13 : 1985 :** NATO Software Quality Control System Requirements

The standard establishes quality control requirements for NATO contractors. It can be used as a stand-alone document or as a supplement to AQAP 1, where software is part of a system using AQAP 1.

The standard consists of a collection of requirements that the contractor must implement to show a Quality Assurance Representative (QAR) that the software quality control requirements are being met.

3.5.3 **AQAP 14 : 1985 :** Guide for the Evaluation of a Contractor's Quality Control system for Compliance with AQAP-13.

This document provides information and guidance to personnel involved in evaluating a contractor's compliance with AQAP-13.

The standard quotes each paragraph of AQAP-13 and then provides guidance and typical questions on the requirements of the paragraph. Some of the "typical questions" checklists are quite extensive.

3.6 RTCA DO-178A :1985: Software Considerations in Airborne Systems and Equipment Certification

This standard deals with software development and quality assurance for avionic software and is in the process of being up-dated to RTCA DO-178B.

4. RECENT/EMERGING STANDARDS

4.1 BS 5750: PART 13:1991/ISO 9000-3: 1991 : Guide to the application of BS5750:Part1 to the development, supply and maintenance of software.

This section of the British Standard was published on 30 September 1991 and forms the basis for the DTI TickIT initiative. This section provides guidance on applying BS 5750, Part 1 / ISO 9001 to the development, supply and maintenance of software.

Under its international title, ISO 9000-3, it will probably become the internationally recognised baseline for software quality issues. It is recommended in IEC 65A for all levels of software integrity.

NB For MoD work, from 1 March 1992, all quality system assessments involving software will be carried out against ISO 9000 and ISO 9000-3, whether carried out by MoD or a third party.

The standard does not address SRCCS directly but provides a framework for applying quality assurance within a software development. The interviews held with the supply side indicate that there is a strong belief within the industry that such a framework should be one of the minimum requirements for any software development.

4.2 IEC Standards

4.2.1 IEC SC45A/WG-A3 (IEC Technical Committee 45 : Nuclear Instrumentation Sub-Committee 45A : Reactor Instrumentation Working Group A3) : Application of Digital Processors

This supplement to IEC 880 [see section 3.3.2 above] is a new standard being developed for software in safety-critical applications of nuclear power stations. The standard concerns new technological developments, the use of formal methods and reuse of existing software and proven components and software diversity. Its scope will be broadened to include safety-related systems in addition to safety-critical systems.

4.2.2 IEC 65A : The International Suite of Standards on Safety-Related Software from the IEC

Several international Working Groups were set up to establish these standards which are still in the review stages.

4.2.2.1 IEC 65A (Secretariat) 122: Version 1: August 1991 : Software for Computers in the Application of Industrial Safety Related Systems

This standard is being developed by Working Group 9 (WG 9). WG 9 dealt with software aspects of the IEC65A suite of standards. This is planned to be the generic standard on safe software and to be taken as a basis for other standards.

This standard is complementary to IEC 65A (Secretariat) 123 (see section 4.2.2.2) and is to be read in conjunction with it. The standard specifies the measures necessary to achieve the software safety requirements identified under IEC 65A (Secretariat) 123. Although primarily concerned with the safety of persons, it is also applicable when serious economic or environmental implications exist.

The standard applies to software used in the implementation and development of safety-related systems including operating systems, support tools and firmware as well as application programming (high/low-level programming and special purpose programming, eg PLC ladder logic and Computer Numerical Controlled Parts programming).

The need to distinguish between safety and reliability aspects is highlighted within this standard. High reliability criteria will require that a different strategy will be adopted from that which would be adopted for high safety criteria. The need to address the safety of the system at the outset and throughout design and implementation is stressed throughout the standard.

The standard defines four software safety integrity levels [IL1-IL4]. The higher the level, the less likelihood there is of a dangerous failure of the PES to be caused by a software specification or design fault. A further non-safety-related level (0) has been included for comparison. Techniques and measures appropriate to all levels are displayed in tables.

The standard requires that an independent assessor shall be appointed to the extent demanded by the integrity level.

The standard does not provide guidance on which level of safety integrity is appropriate for a given risk, since this decision will depend upon many factors, including the nature of the application, the extent to which other systems carry out safety functions and social and economic factors.

Principles applied in developing high integrity software include:

- top-down design methods
- modularity
- verification of each phase of the development lifecycle
- verified modules and module libraries
- clear documentation
- auditable documents
- validation testing.

The different safety integrity levels in this standard require different levels of assurance that these and related principles have been correctly applied.

4.2.2.2 IEC 65A (Secretariat) 123: Version 1: September 1991: Functional Safety of Programmable Electronic Systems : Generic Aspects

This standard is being developed by Working Group 10 (WG 10). WG 10 has been mainly concerned with the system aspects of the IEC65A suite of standards. This provides a basis for the IEC 65A (Secretariat)122 standard (see section 4.2.2.1 above) and provides guidelines on the aspects that need to be addressed when programmable electronic systems (PESs) are used to carry out safety functions.

The standard provides a classification of software integrity levels and the requirements for the system requirements specification, design and validation. The standard therefore addresses the specification of the safety functions to be allocated to the software in the system.

The approach to developing the standard has been that of a systematic approach to:

1. identifying hazards, risks and risk criteria

2. identifying the necessary risk reduction to meet the risk criteria

3. defining an overall Safety Requirements Specification for the safeguards necessary to achieve the required risk reduction

4. planning, monitoring and controlling the technical and managerial activities necessary to translate the Specification into a Safety-Related System of a validated safety performance (or safety integrity).

4.2.3 ISO/IEC 9126 : 1991 : Information Technology - Software product evaluation - quality characteristics and guidelines for their use

The standard provides a definition of terms for assessing the quality of a software product. It establishes a set of quality characteristics (metrics) that describe the product and hence provides a potential basis for a quantitative evaluation of software quality.

4.3 MoD Standards

4.3.1 DEFSTAN 00-22 : Issue 2 : May 1991 : The Identification and Marking of Programmable Items

The standard provides a coherent scheme for the marking of firmware ie where software is embodied in a Programmable Item (PI). The standard specifically addresses cases where such items are to be reprogrammed and consequently reidentified in service. This is necessary because the physical appearance of the equipment may not change but the function it performs may if it is reprogrammed. This will have obvious safety implications.

4.3.2 **DEFSTAN 00-38 : INTERIM STANDARD : June 1991 :** Guidelines for the Evaluation of Microprocessors for Avionics Applications

The standard provides project managers, suppliers and designers of avionics equipment with broad guidance on selection criteria for electronic microprocessors and advice on applying the criteria in order to aid in the selection of microprocessors for avionics applications.

The standard does not relate specifically to safety-critical applications and does not address system issues. The standard identifies and defines three classes of selection criteria: general, operational and performance.

The items and procedures in the standard may be claimed to be subject to patent/copyright/design rights in the UK or other countries.

4.3.3 **DEFSTAN 00-55 : INTERIM STANDARD : 1991 :** The procurement of safety-critical software in defence equipment

The standard was produced by the Ministry of Defence and is based on the MoD procurement lifecycle.

Part 1 of the standard states the requirements. Part 2 contains guidance on the requirements of Part 1 in order to provide technical background and to elaborate on the requirements so that conformance is easier to achieve and assess.

The standard specifies requirements for all software used in Defence Equipment designated as Safety Critical but it may also be applied in whole or in part to non-safety-critical software. It does not deal with the safety of the whole system (which is addressed in DEFSTAN 00-56) but sets out procedures for the software development process which are required over and above those set out in DEFSTAN 00-16 and DEFSTAN 00-31 for software of lower integrity levels.

The requirements of the standard extend to tools and support software used to develop, test, certify and maintain safety-critical systems through all phases of the project lifecycle, if hazard analysis shows such tools to be safety-critical.

The standard addresses software integrity levels. The standard is aimed not only at the use of formal methods but at:

- the use of existing software

- gaining a better understanding of the system during the specification and design stages by applying static path analysis and dynamic testing.

The standard also requires:

- the appointment of an Independent Safety Auditor from the outset of the project, under a separate contract.
- the development, at the design stage, of a Safety Plan which will show the detailed safety planning and control measures that will be employed.
- formal Safety Reviews to be carried out by a Review Committee
- the production of a Code of Design Practice
- the maintenance of a Safety Records Log throughout the entire lifecycle of the software (ie including in-service phase)
- documentation for the SCS that covers all lifecycle phases and that is sufficiently comprehensive for the entire SCS to be re-established. The documentation shall be produced in accordance with JSP 188.
- configuration management as specified in Annex C of the standard
- certification and acceptance into service (the proforma for the certificate is given in Annex D of the standard)
- renegotiation of liabilities covered by the Safety Critical Software Certificate if any part of the SCS is transferred to a third party.
- that the disposal and decommissioning of the SCS be treated as a project in their own right.

A list of deliverable items is given in Annex B of the standard.

A warning is given within Part 1, Section 1.2 of the standard that the use of the procedures, techniques practices and tools referred to in the standard is expected to provide greater assurance that software is free from errors than if they were not used. This does not absolve the designer, producer, supplier or user of the system from statutory obligations relating to safety at any stage.

The standard represents a significant improvement in MoD procedures by replacing subjective assessment with rigorous disciplines.

4.3.4 DEFSTAN 00-56 : INTERIM STANDARD : 1991 : Requirements for the analysis of safety critical hazards for computer based systems

This standard is not aimed just at software but at a standardized approach to hazard analysis of the whole system, including its interface with its operational environment. The standard requires documentary evidence that safety management has been properly applied throughout the lifecycle of the system, ie from feasibility phase through to removal from service. The various hazard analysis activities are to be recorded in a Hazard Log for the system which is to be maintained throughout the project lifecycle.

The standard sets out the techniques and procedures for carrying out and recording hazard analysis and safety risk assessment on new systems, and systems undergoing modification or maintenance. The objective of the procedures is to identify, evaluate and record the hazards of the system so that the maximum tolerable risk can be determined and to facilitate the achievement of a risk that is as low as reasonably practicable (ALARP) and below the maximum tolerable level. The standard states that this activity will provide the safety criteria to arrive at a reasonable and acceptable balance between the reduction of the risk to safety and the cost of that risk reduction.

The standard also establishes a safety classification to be applied to defence systems. Safety is concerned with the potential to endanger human life.

4.3.5 DEFSTAN 05-61 : PART 15 : ISSUE 1 (THIRD DRAFT) : January 1992: Quality Assurance Procedural Requirements Part 15 : Arrangements for Installation of Fixed Ground Based Equipment Incorporating Electronic Systems

The standard provides procedures to be implemented at the time and place of installation of fixed ground-based equipment incorporating electronic systems. Emphasis is placed on there being the same quality control for site work as "at the works" , eg arising from amendments to software, drawings and specifications, to installation software/hardware .

4.3.6 DEFSTAN 05-95 : INTERIM STANDARD : September 1991 : Quality System requirements for the Development, Supply and Maintenance of Software

This interim standard is provisional because it has not previously been circulated for public comment. It is composed of requirements from ISO 9001, text from ISO 9000-3 which has been modified to express a requirement (since ISO 9000-3 is a guidance document) and supplementary requirements from DEFSTAN 05-91 and AQAP-13.

5. REFERENCES

1 Ehrenberger W., "Report on Connections to Standardisation Bodies from September 1990 to August 1991", Esprit SCOPE Project Document No. SC.91/002/Ehr/R 1.1.4, Sept. 1991

APPENDIX H

THE GERMAN TÜV SYSTEM FOR ACCREDITATION OF SAFETY RELATED COMPUTER CONTROLLED SYSTEMS

PREPARED BY

SRD (AEA TECHNOLOGY)

The German TÜV System for Accreditation of Safety Related Computer Controlled Systems

1. INTRODUCTION

Several of those interviewed on the supply and demand sides mentioned the German TÜV accreditation system for safety-related computer-controlled systems (SRCCS). This section is designed to provide information on the TÜV system. It must be noted that information in English on the TÜV system is not easy to find.

Information presented below on the TÜV system and its application has been obtained from a system retailer who is involved with suppliers of German components for which TÜV accreditation has been applied for and granted. This route to the information has helped identify the practicalities involved in seeking such accreditation.

The TÜV system has developed over a period of years from being based upon a series of 'black box' generic tests to including static analysis and review of the process [ref "European Safety Testing: The Role of Static Path Analysis by R Faller MILCOMP 1992].

Standards similar to IEC 65A are being developed in Germany to feed into the rating system for the SRCCS.

2. THE APPROVAL PROCESS

To illustrate the process, a PLC is used as an example of computer-controlled equipment subjected for accreditation.

The following are the main features of the approval process:

- the approval process determines, by physical testing, whether the safety features of the PLC concerned are commensurate with a specified level of safety determined by the application (ie consequences of failure). The process **does not** analyze the safety of the equipment in the application, and hence dangerous failure modes of the system will not be detected by this form of approval.

- the approval involves taking a production model of the PLC concerned, and subjecting it to a series of tests by TÜV on their test rigs. The range and rigour of these tests depend upon the consequences of failure of the PLC, and hence a safety classification system has been evolved to cater for the consequences.

- the system of safety classes on which approval is granted is driven by the requirements of existing (German) regulations.

- manufacturers apply for accreditation against a safety category or safety categories

- testing is conducted against the requirements of relevant standards.

- the classification scheme consists of safety categories 1 to 5, with category 1 being the highest [ref "Microcomputers in Safety Technique - an Aid to Orientation for

Developer and Manufacturer", by Hölscher & J Rader, Verlag TÜV Bayern, Rheinland, GmbH, Köln, ISBN 3-8858-315-9]

- the approval scheme involves subjecting the system to a number of tests, including environmental tests and functional hardware and software tests, which involves, for example, a demonstration of safe system shutdown under failure conditions.

- on satisfactory completion of the testing a Test Certificate is issued to the manufacturer which details the:

    - Test Object, ie what was tested

    - Type designation, ie model number

    - Manufacturer

    - Application, eg safety-related use in plants of Safety Category 2

    - Base of tests, ie the standards used as the basis for the tests,

    - Test Result, ie which sections of the software were tested (eg the logical functions, the entire operating system) and whether all tests were passed

    - Special conditions, eg any special conditions which may apply when the system is installed in plants of certain safety categories.

- TÜV approval takes of the order of six to nine months to obtain.

3. SAFETY CATEGORIES

Safety levels are defined on the basis of the number of faults which must combine to produce a failure of the system and are comparable with those in IEC 65A.

As well as the general exclusion of dangerous primary defects, measures are also required for failure detection to prevent the dangerous breakdown within a preset time if the following fault combinations occur:

- **Safety Category 1**

    The combination of any faults must not cause a dangerous breakdown, even if faults occur in any time sequence and are not detected by the regular test routines.

    Typical standards on which accreditation might be based are:

    DIN/VDE 0831 - Railway Signal Plants

- **Safety Category 2**

    The combination of up to three undetected, individually non-dangerous faults must not cause a dangerous breakdown.

Typical standards on which accreditation might be based are:

TRA 200/101 - Elevator Controls

- **Safety Category 3**

  The combination of up to two individually non-dangerous faults must not cause a dangerous breakdown.

  Typical standards on which accreditation might be based are:

  DIN/VDE 0116 - Electrical Equipment for Furnaces

- **Safety Category 4**

  Any single fault must not cause a dangerous breakdown.

  Typical standards on which accreditation might be based are:

  DIN/VDE 0113 - Handling Machines

- **Safety Category 5**

  Defined types of faults must not cause a dangerous breakdown.

4. Definition of Risk in DIN V 19250

Risk has been defined in the Proposed DIN V 19250 standard, "Fundamental Safety Aspects to be Considered for Measurement and Control Equipment" in the following way:

Risk (R) = probability of frequency (F) and level of harm (H)

$R = F \times H$ and

$F = f(T,D,P)$ where:

T = Time persons staying in the dangerous area
D = Danger prevention
P = Probability of the unwanted event

4.1 Risk parameters

4.1.1

H	**Level of Harm**
H1	small injury
H2	irreversible injury of one or more persons, death of one person
H3	death of some persons
H4	catastrophe, disaster, death of many persons
T	**Time people stay in the dangerous area**
T1	very seldom or sometimes
T2	frequently or permanent
D	**Danger prevention**
D1	possible, depending on conditions
D2	not possible
P	**Probability of unwanted event**
P1	very small
P2	small
P3	high

4.2 Relationship of the Risk Parameters

The following chart presents the relationship between the risk parameters as proposed in DIN V 19250.

<table>
<tr><td rowspan="10">R</td><td colspan="3"></td><td>P3</td><td>P2</td><td>P1</td></tr>
<tr><td colspan="3">H1</td><td>1</td><td>-</td><td>-</td></tr>
<tr><td rowspan="4">H2</td><td rowspan="2">T1</td><td>D1</td><td>2</td><td>1</td><td>-</td></tr>
<tr><td>D2</td><td>3</td><td>2</td><td>1</td></tr>
<tr><td rowspan="2">T2</td><td>D1</td><td>4</td><td>3</td><td>2</td></tr>
<tr><td>D2</td><td>5</td><td>4</td><td>3</td></tr>
<tr><td rowspan="2">H3</td><td colspan="2">T1</td><td>6</td><td>5</td><td>4</td></tr>
<tr><td colspan="2">T2</td><td>7</td><td>6</td><td>5</td></tr>
<tr><td colspan="3">H4</td><td>8</td><td>7</td><td>6</td></tr>
</table>

Table 1: The Risk Graph

The risk graph leads to 8 requirement categories, such that:

1 = lowest requirement category

8 = highest requirement category.

eg to calculate the requirement category of a burner control system, the category is obtained using the following criteria:

H3 death of some people

T1 people stay seldom or sometimes in the dangerous area

danger prevention not possible

P2 probability of an unwanted event is very small

Hence H3 → T1 → P2 results in requirement category 5.

These new requirements categories are related to the TÜV safety categories in the following way:

<table>
<tr><th>requirement category
(DIN V 19250)</th><th>safety category
(handbook of TÜV)</th></tr>
<tr><td>8</td><td></td></tr>
<tr><td>7</td><td>1</td></tr>
<tr><td>6</td><td>2</td></tr>
<tr><td>5</td><td>3</td></tr>
<tr><td>4</td><td>4</td></tr>
<tr><td>3</td><td rowspan="2">5</td></tr>
<tr><td>2</td></tr>
<tr><td>1</td><td></td></tr>
</table>

Table 2 : Requirement/Safety Category Relationship

At both ends of the scale, DIN V 19250 has one category more:

- in requirement category 1 the requirements are very poor.
- in requirement categories 7 and 8 the requirements are mainly satisfied by organized measures.

Therefore there are no **essential** discrepancies between DIN V 19250 and the handbook of the TÜV.

5. Measures to attain Requirement Categories

Measures to attain requirement categories, ie to reduce the risk to tolerable levels, are also defined and are similar to those described in IEC 65A.

6. Conclusions

The TÜV system provides a certificate to show that the system has passed a set of dynamic tests which are geared to the safety category of the plant on which the system will reside.

The certificate does not assess the quality of the process which developed the SRCCS or indicate dangerous failure modes of the system.

Information on the system is difficult to obtain but a few UK companies are aware of the system and use it.

Printed in the United Kingdom for HMSO
Dd294941 C17 10/92 G3397 10170